LEONARDO DA VINCI'S CODEX LEICESTER | *A New Edition*

LEONARDO DA VINCI'S CODEX LEICESTER

LEONARDO DA VINCI'S CODEX LEICESTER *A New Edition*

DOMENICO LAURENZA & MARTIN KEMP

THE CODEX

VOLUME I

OXFORD
UNIVERSITY PRESS

THE CODEX LEICESTER

PART 1

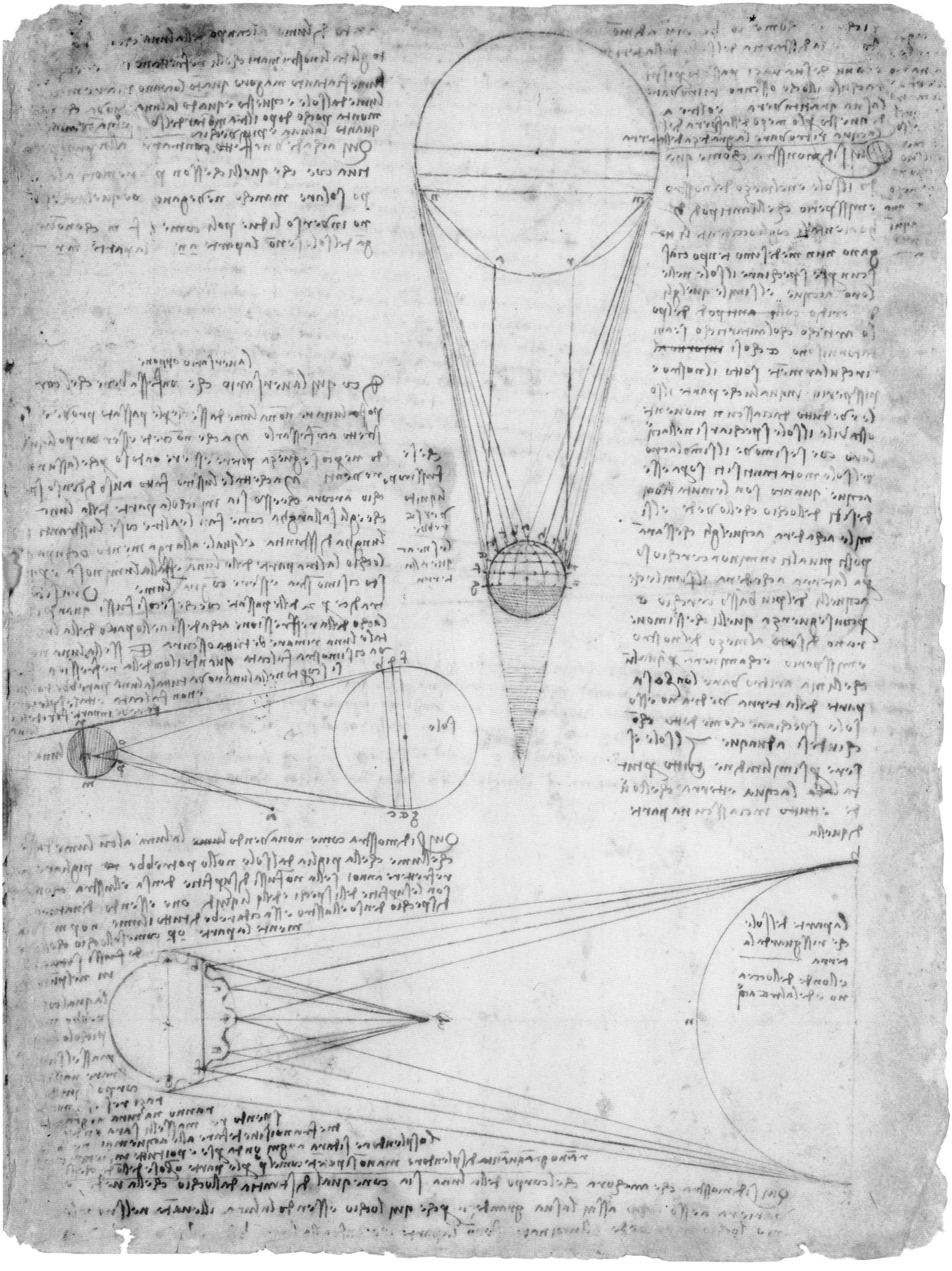

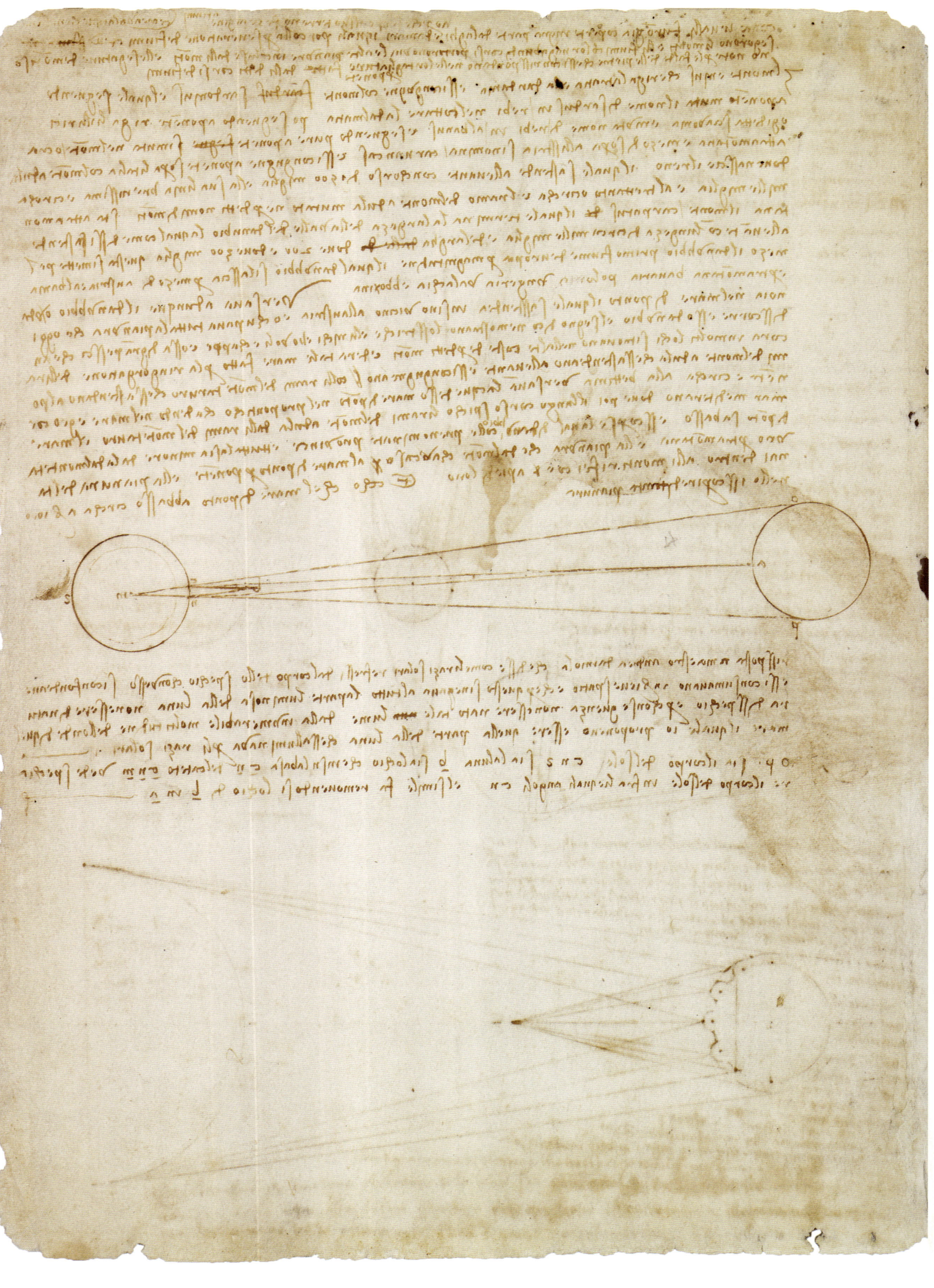

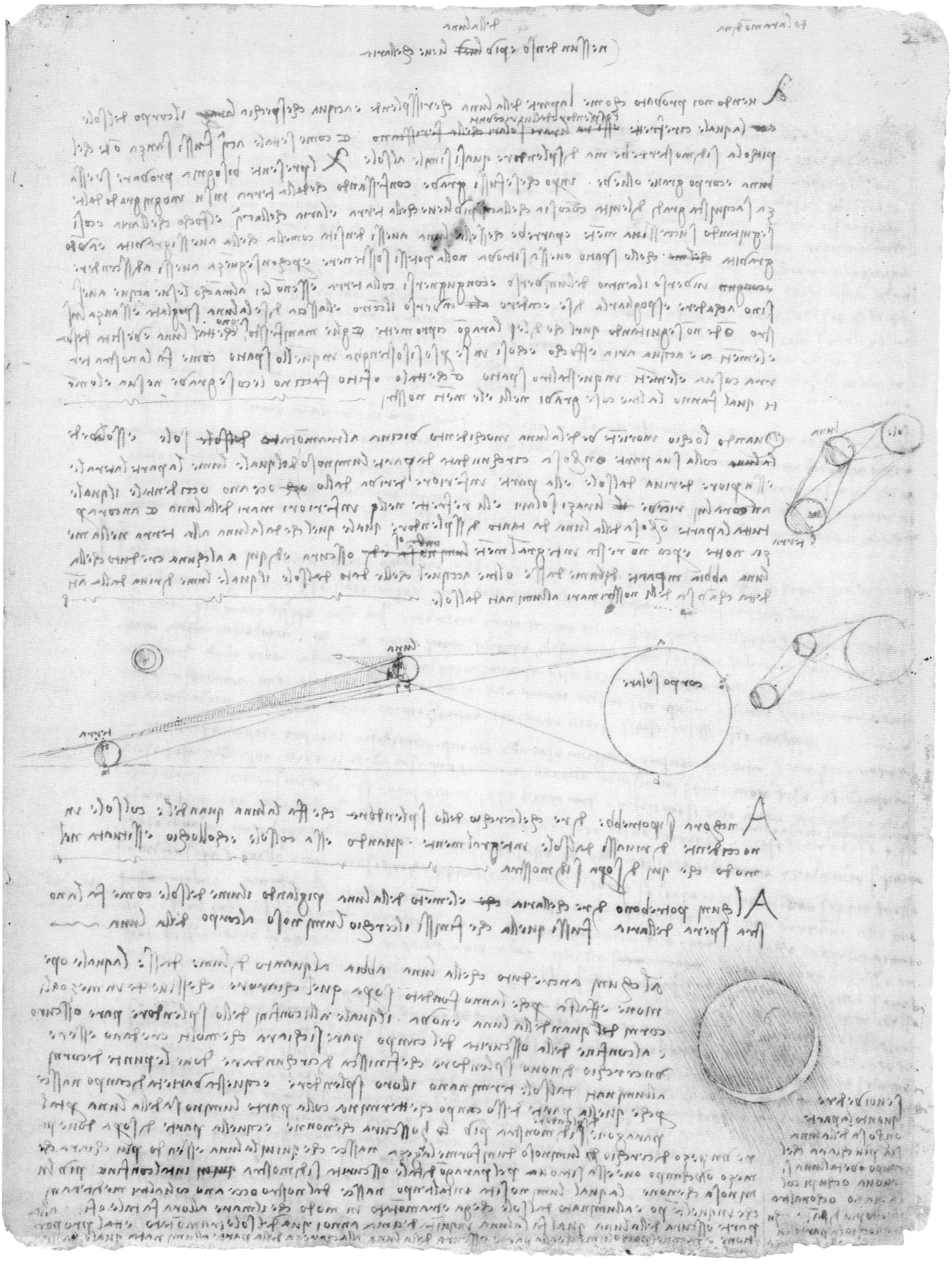

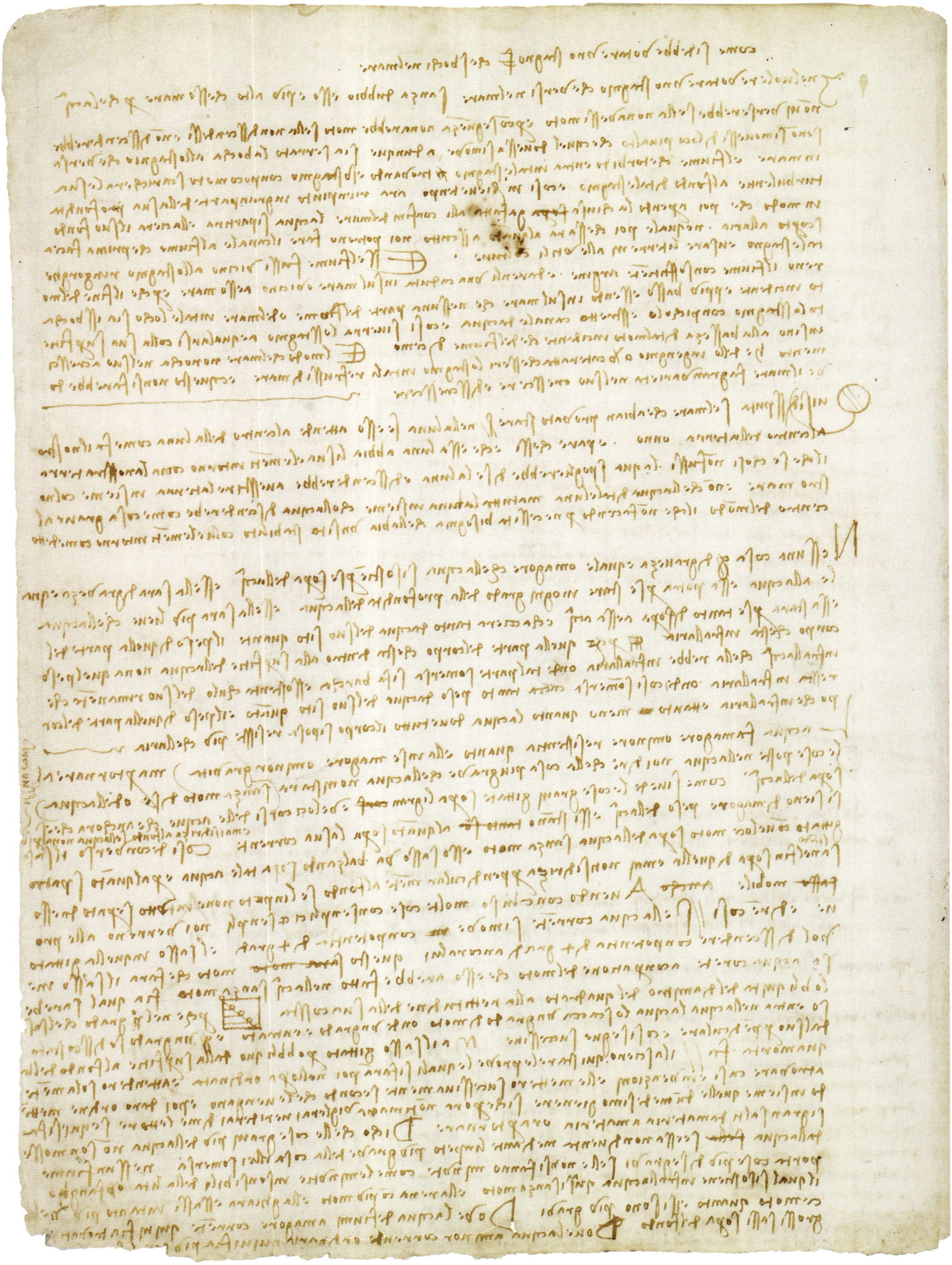

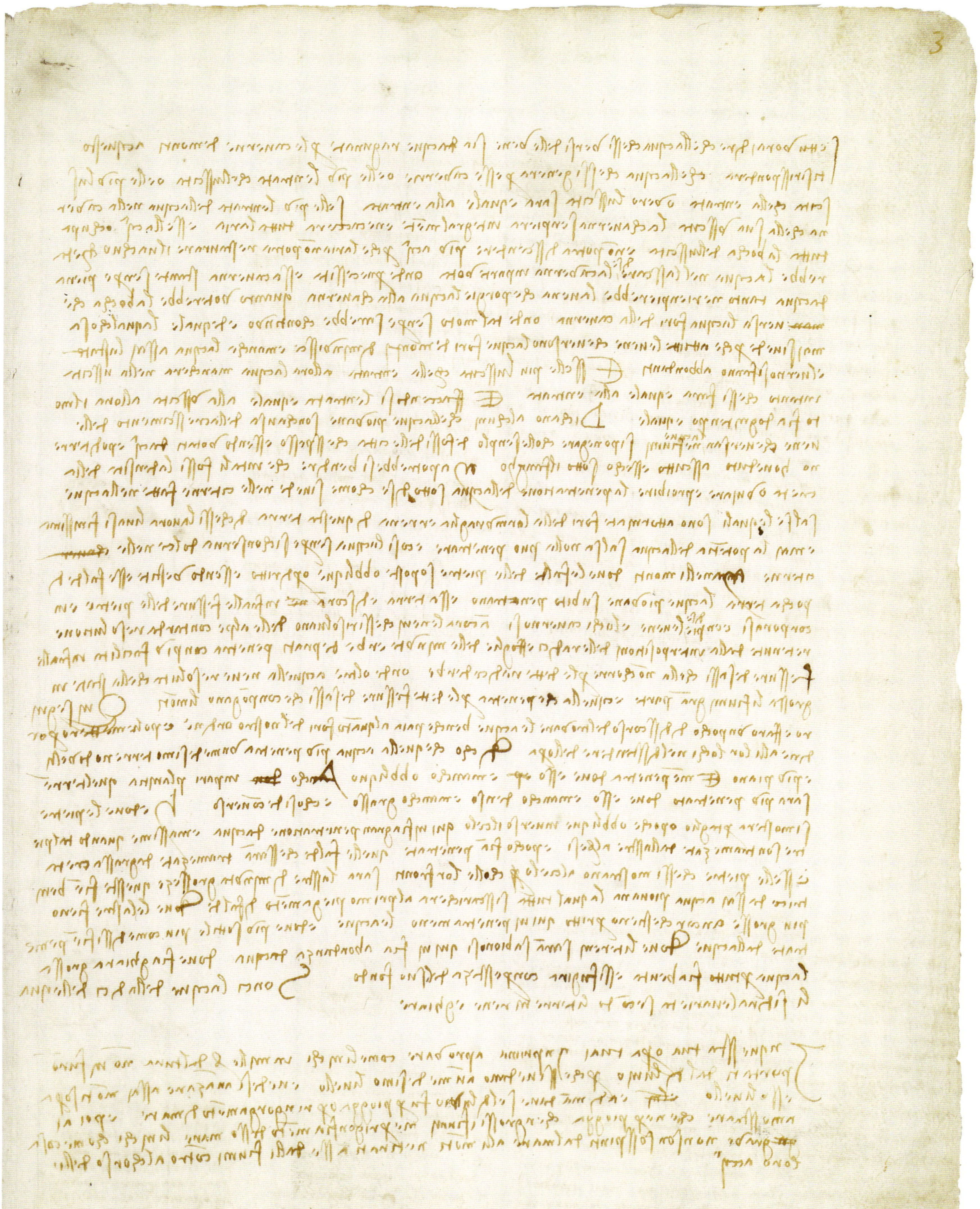

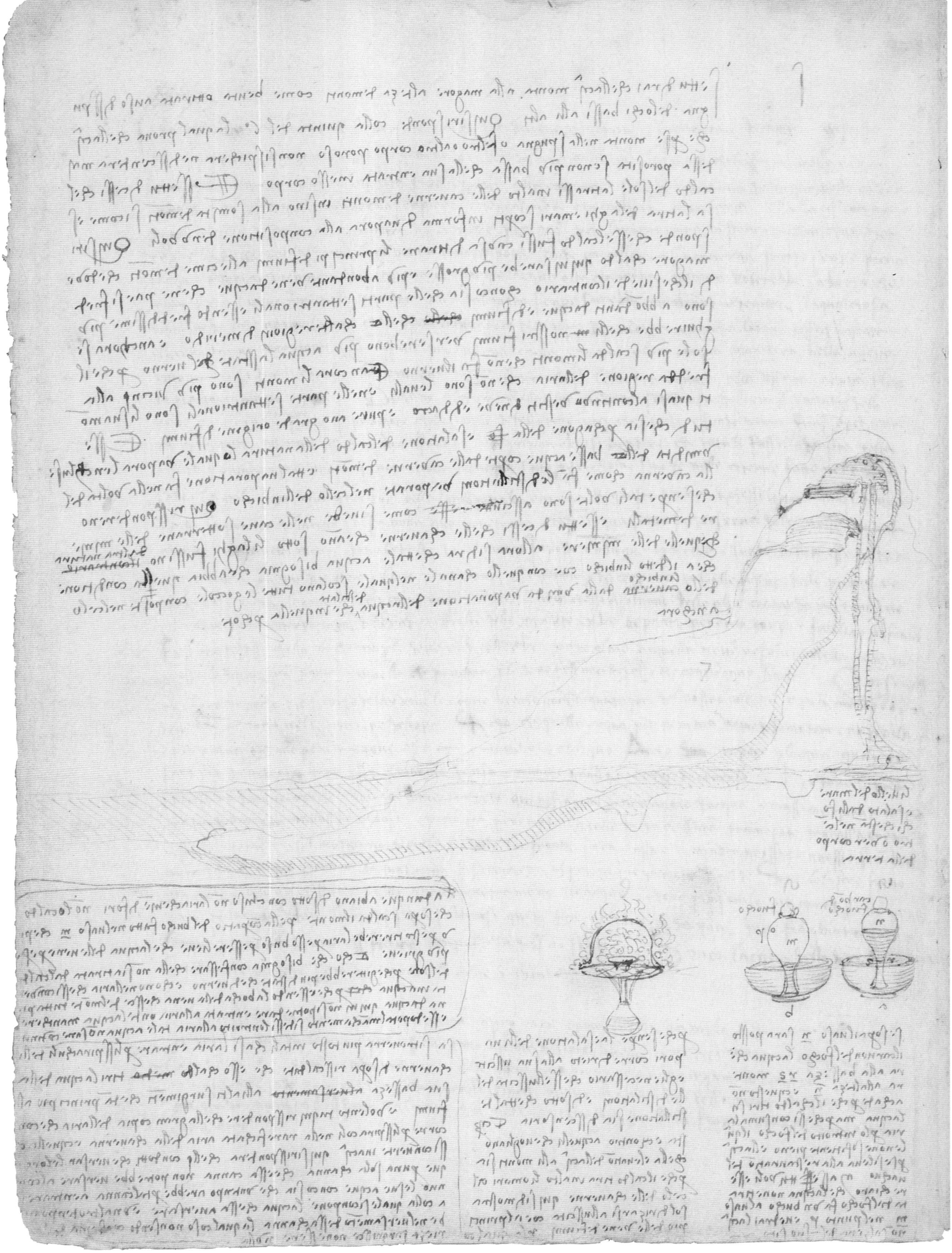

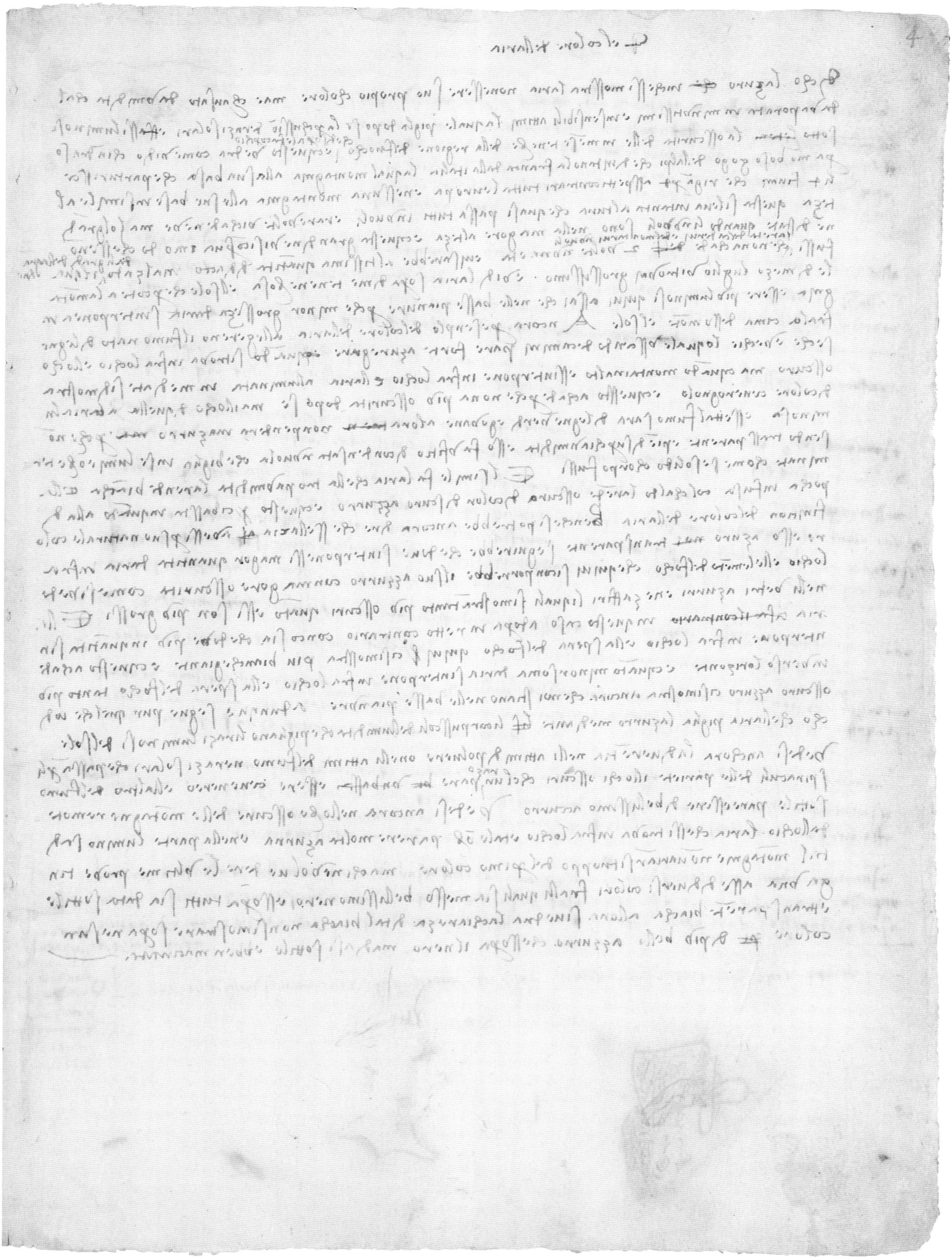

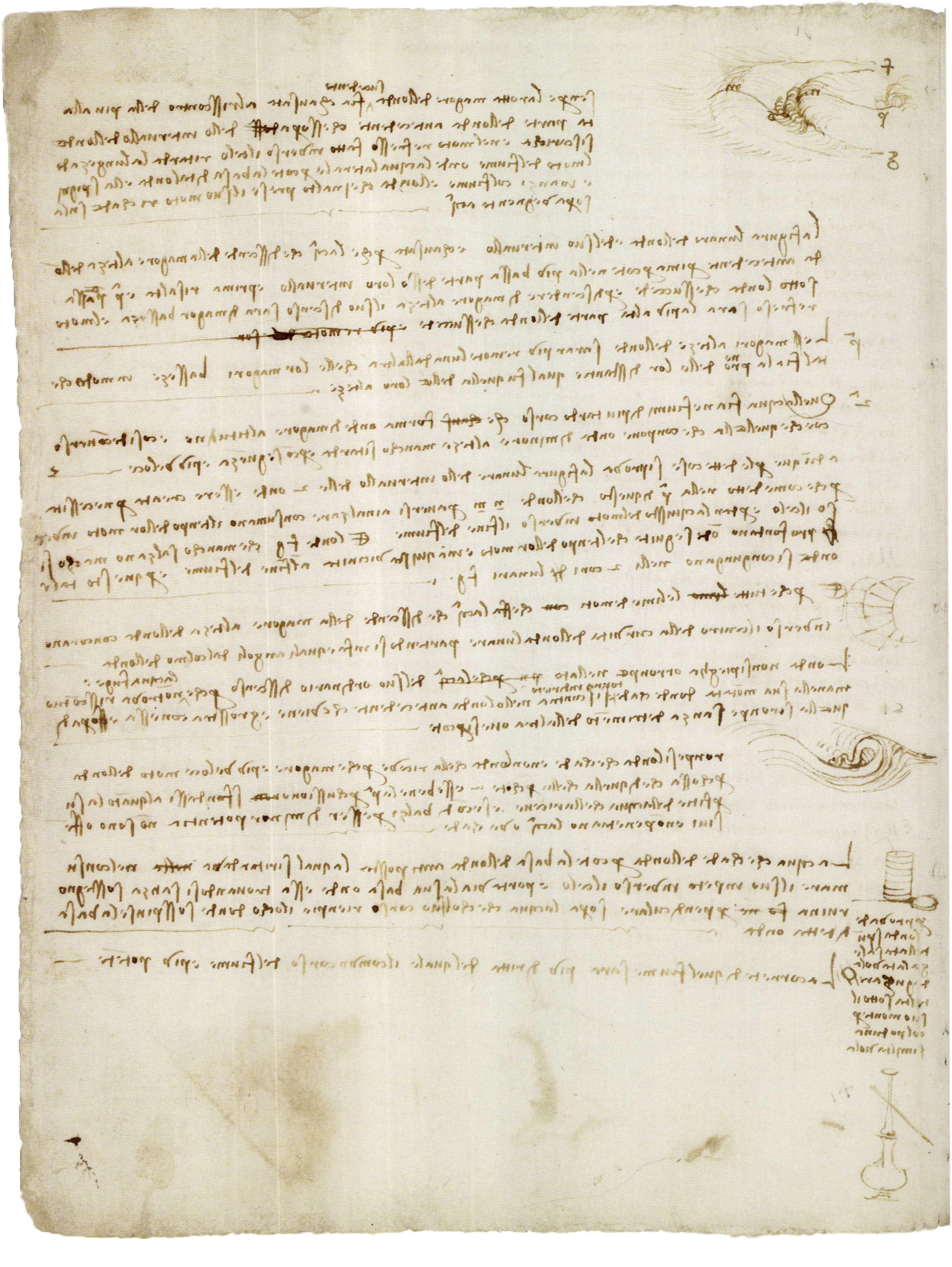

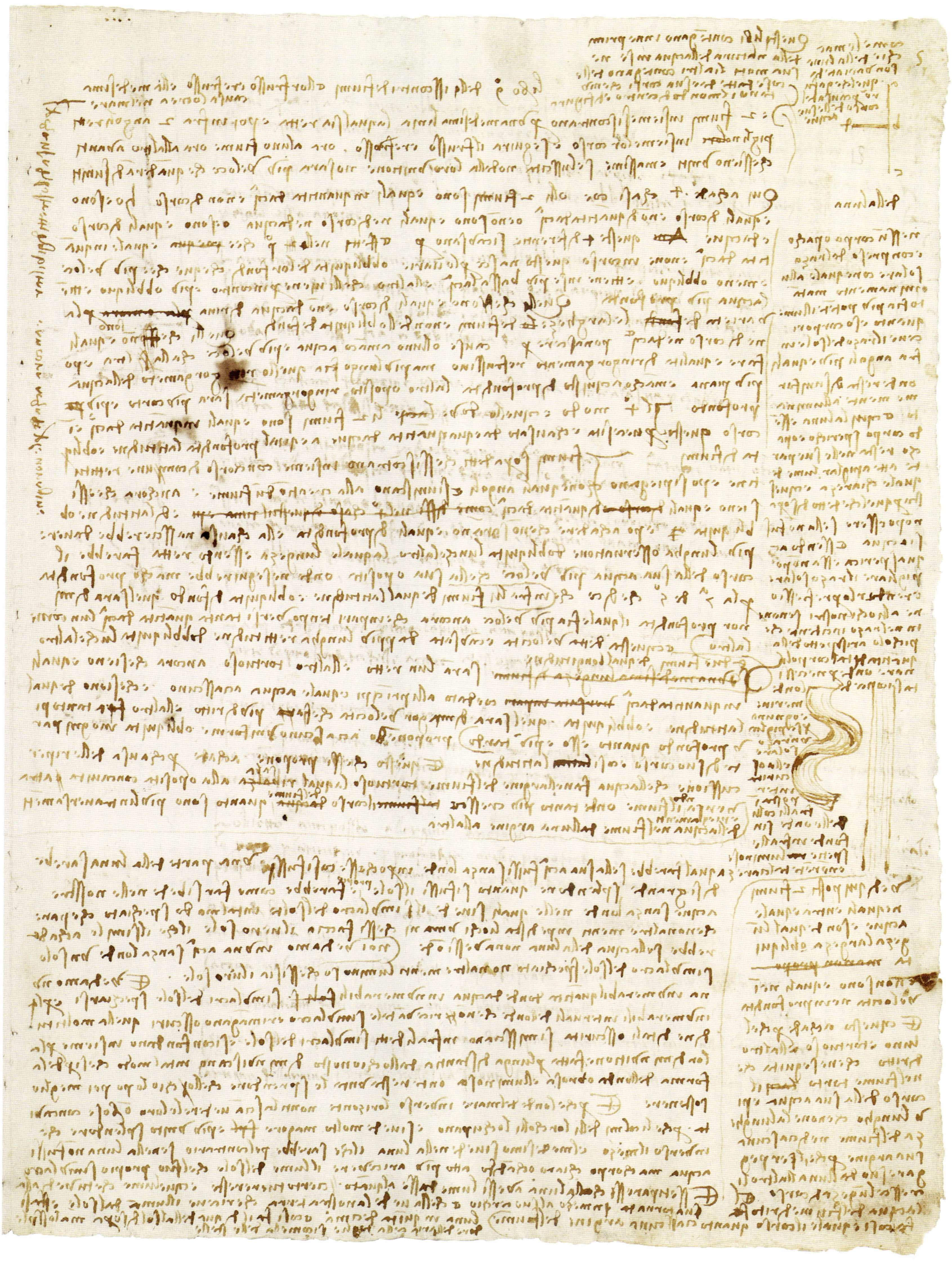

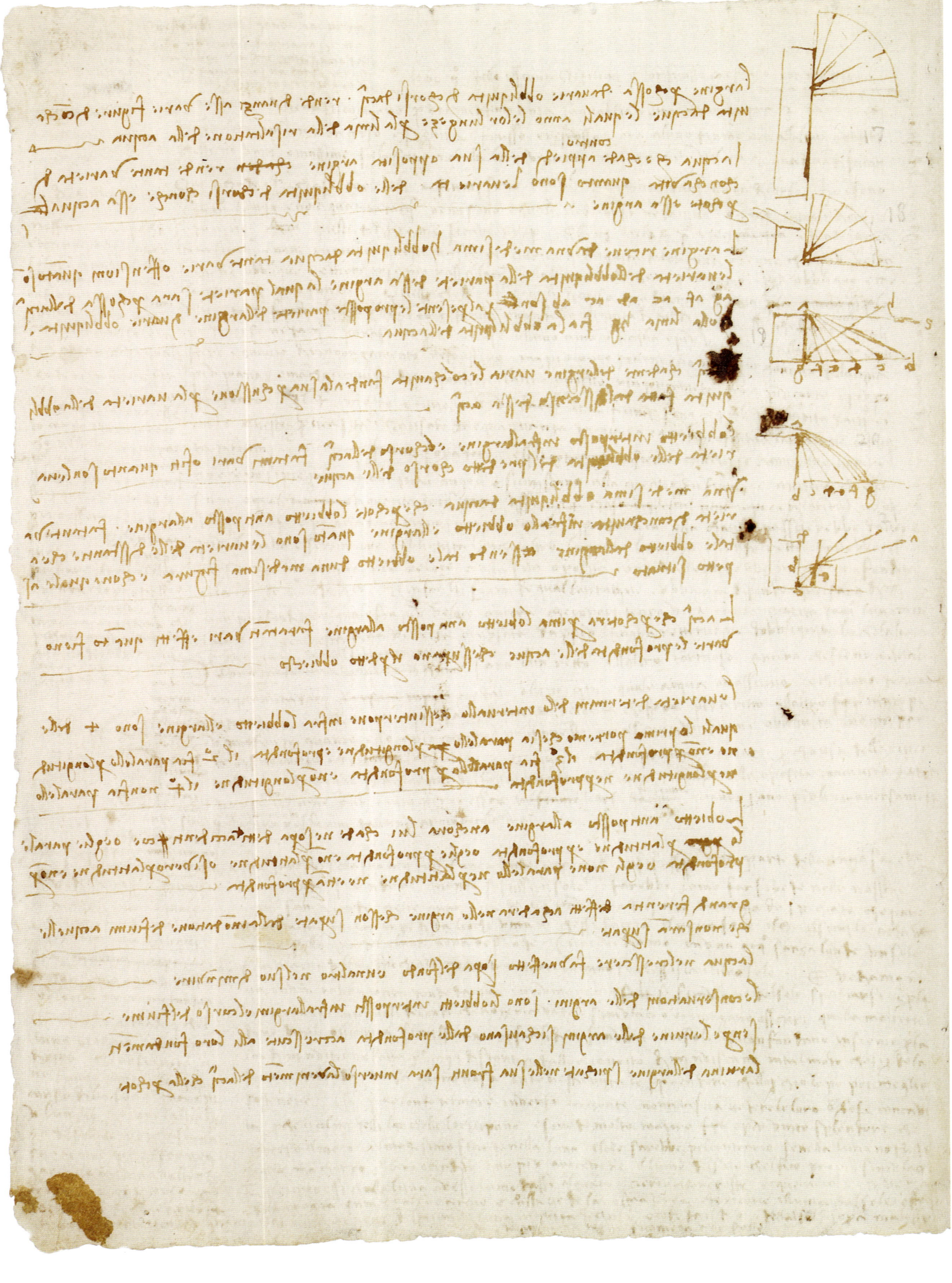

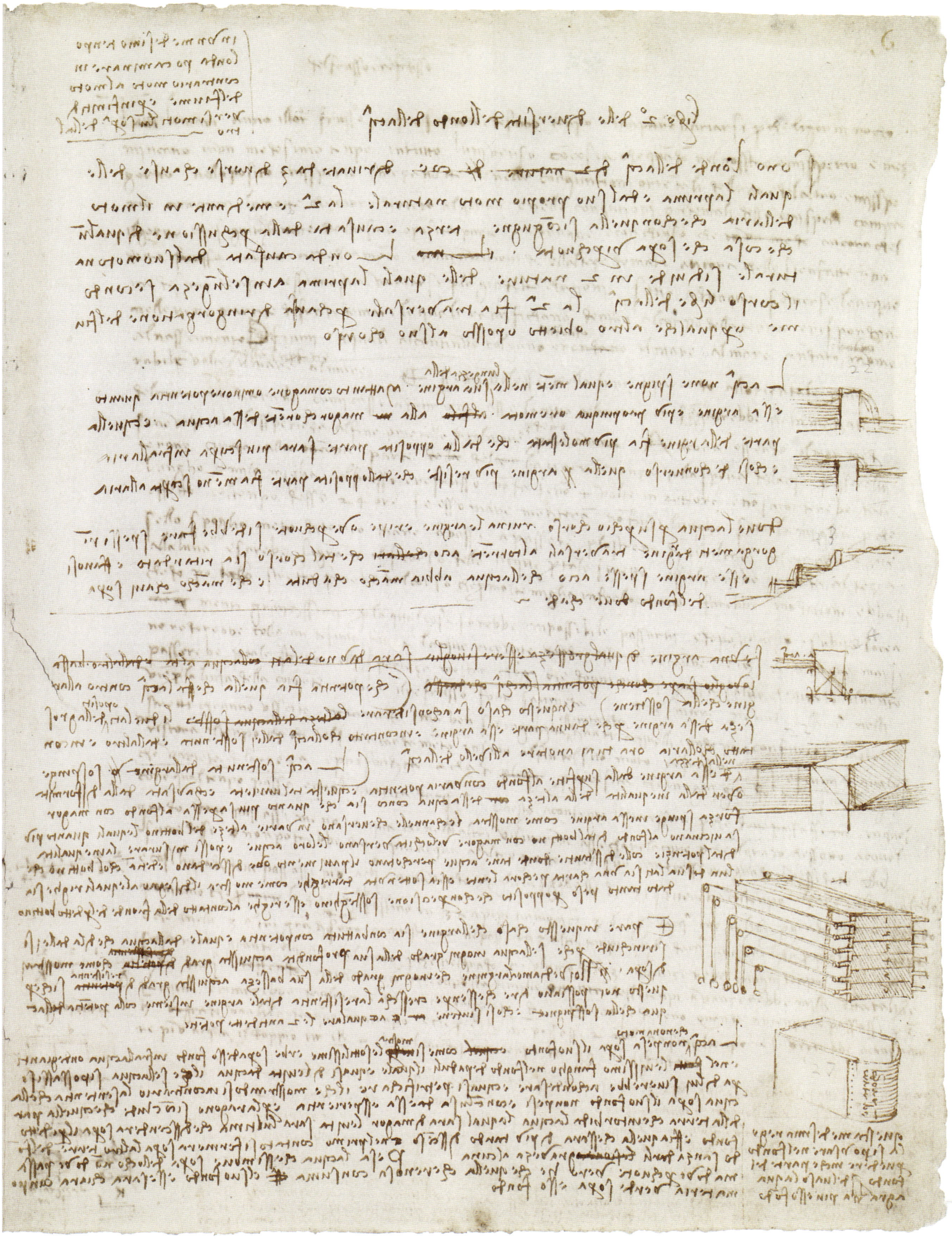

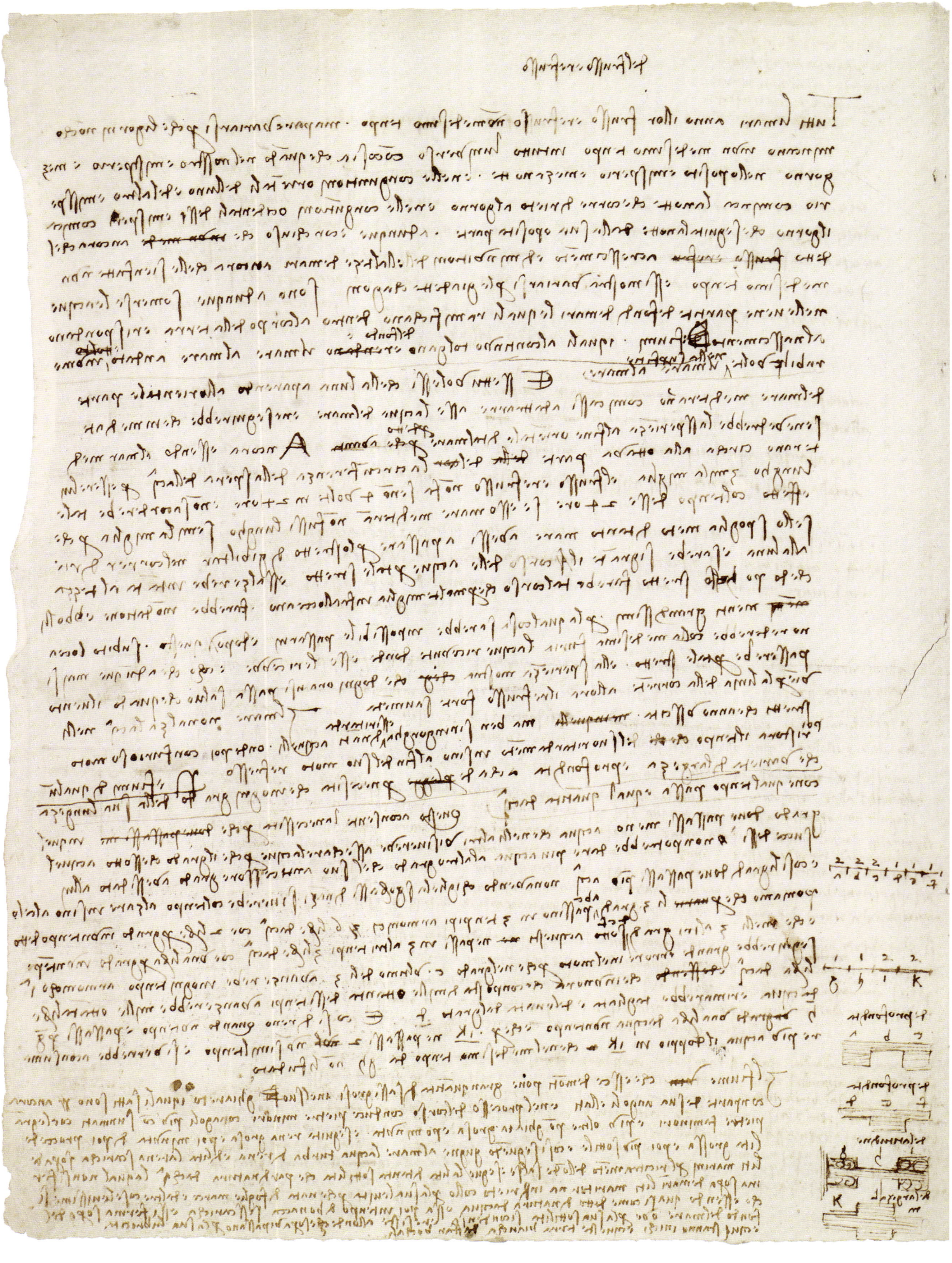

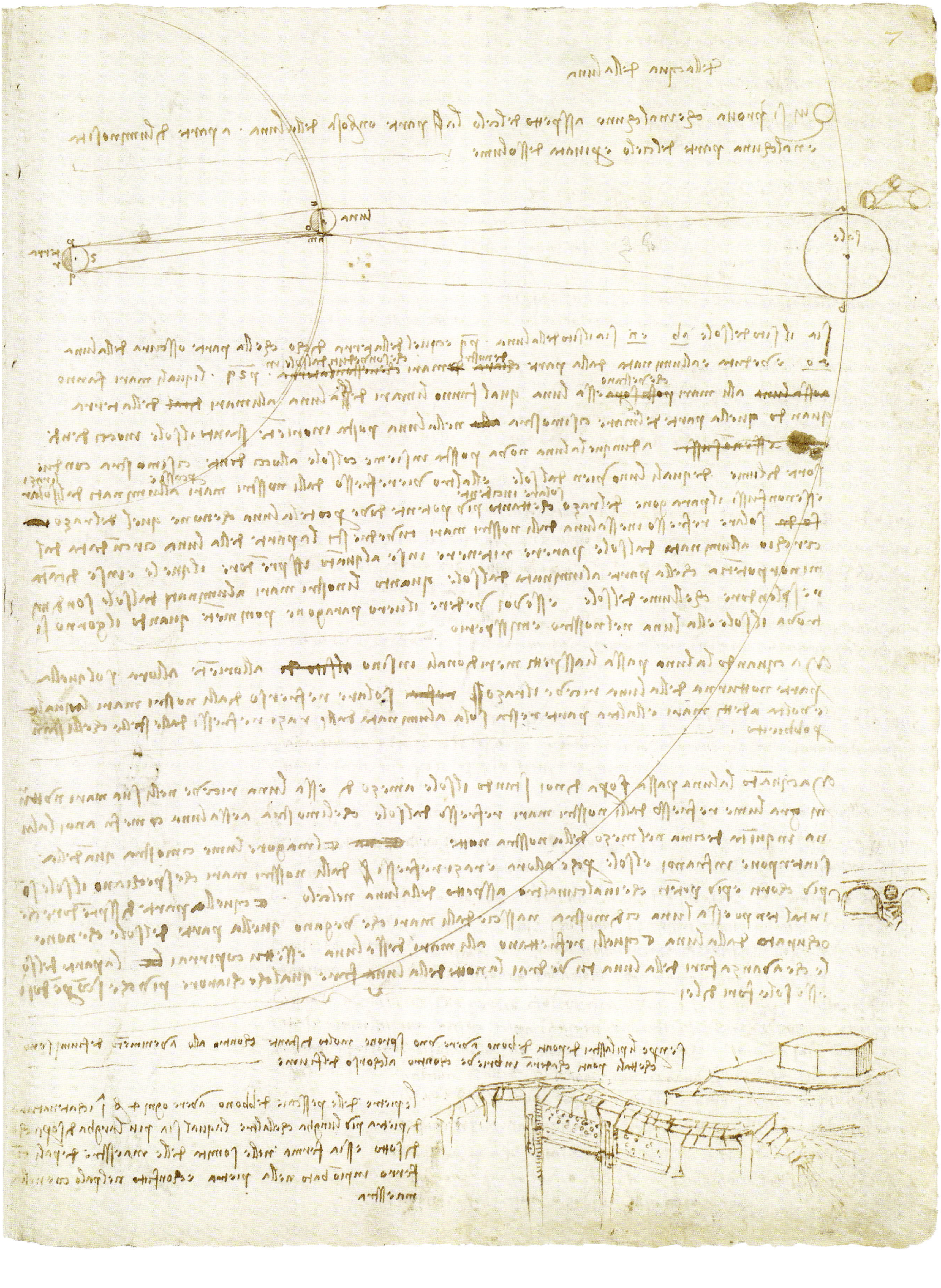

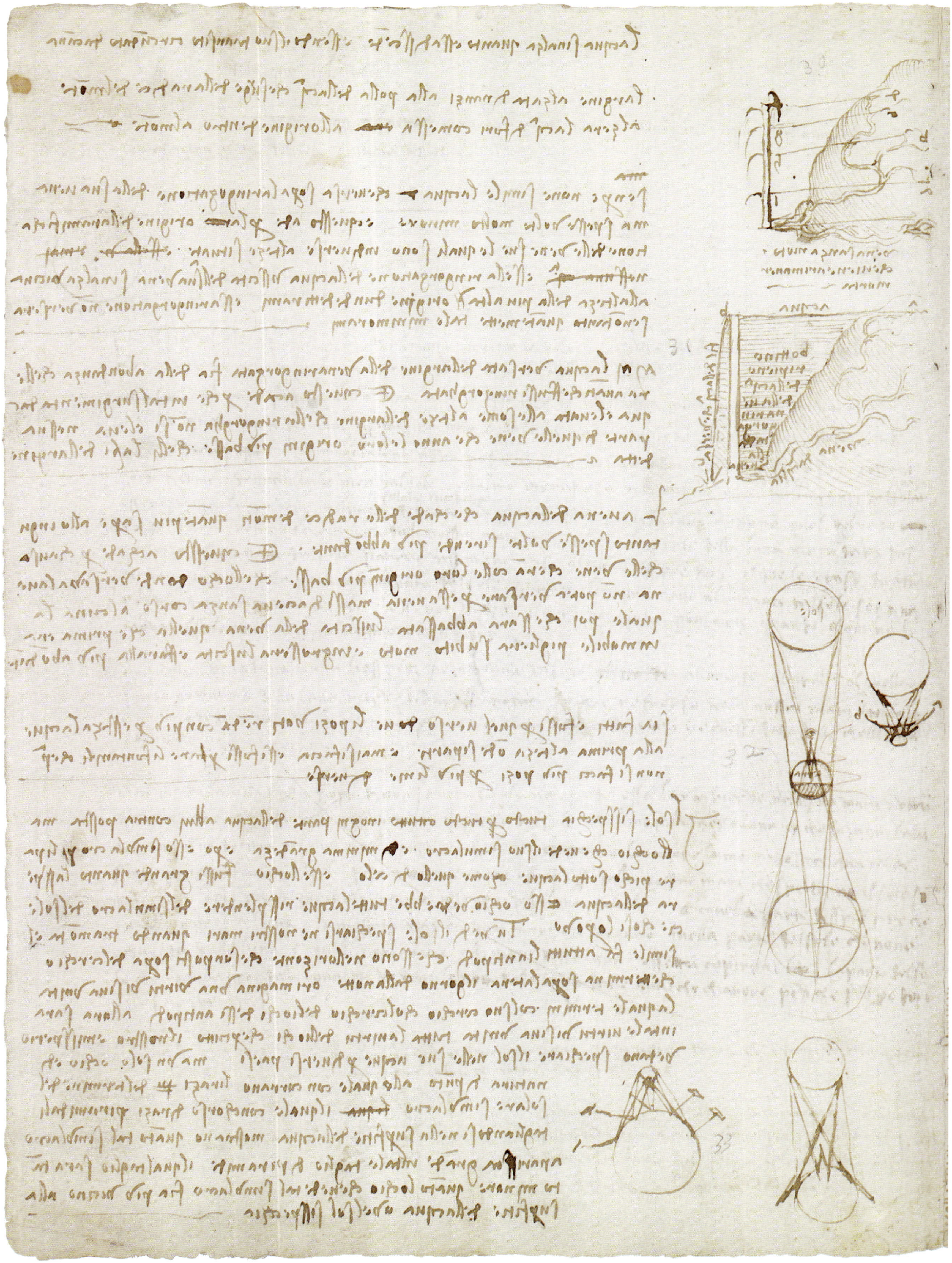

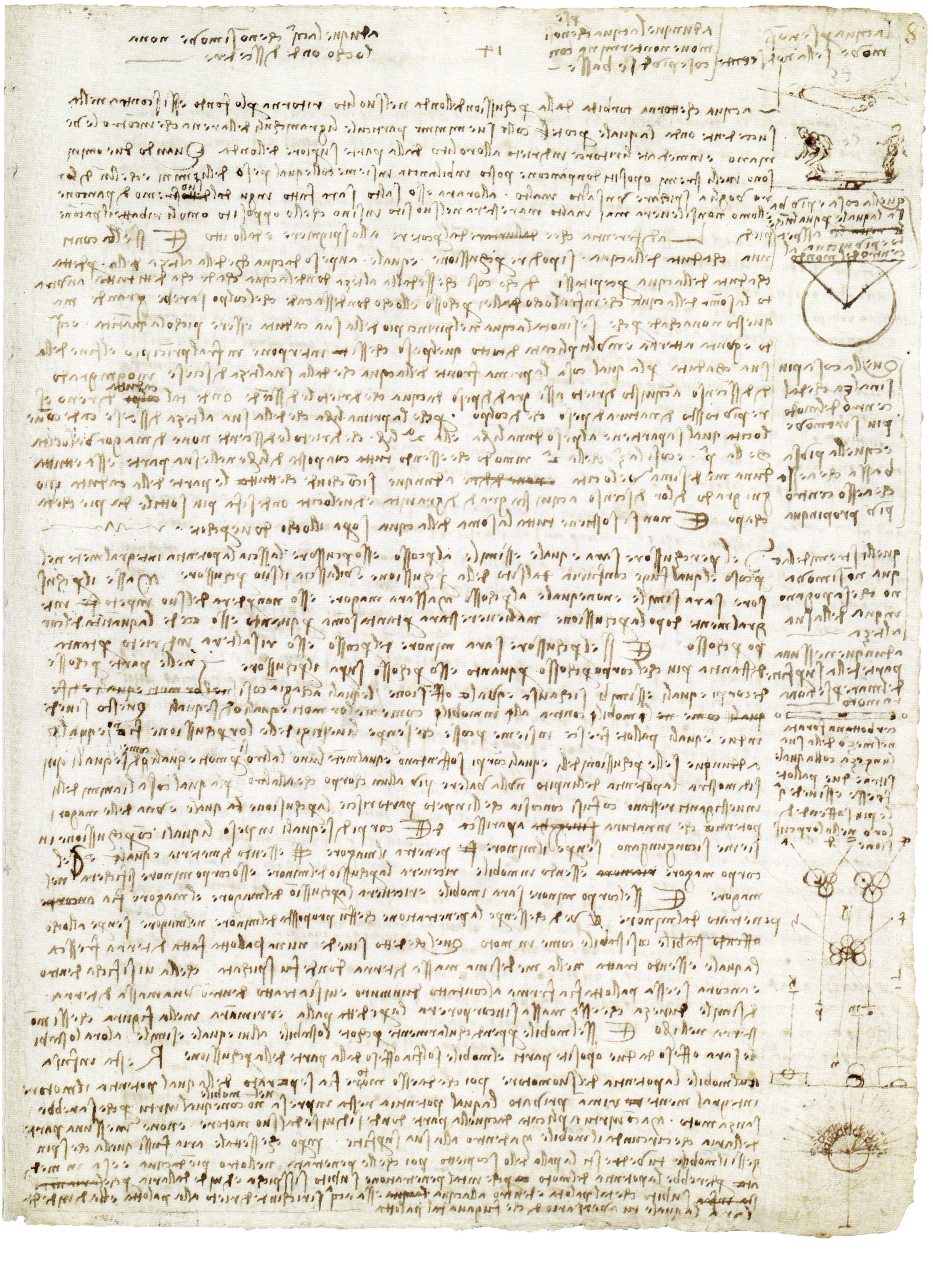

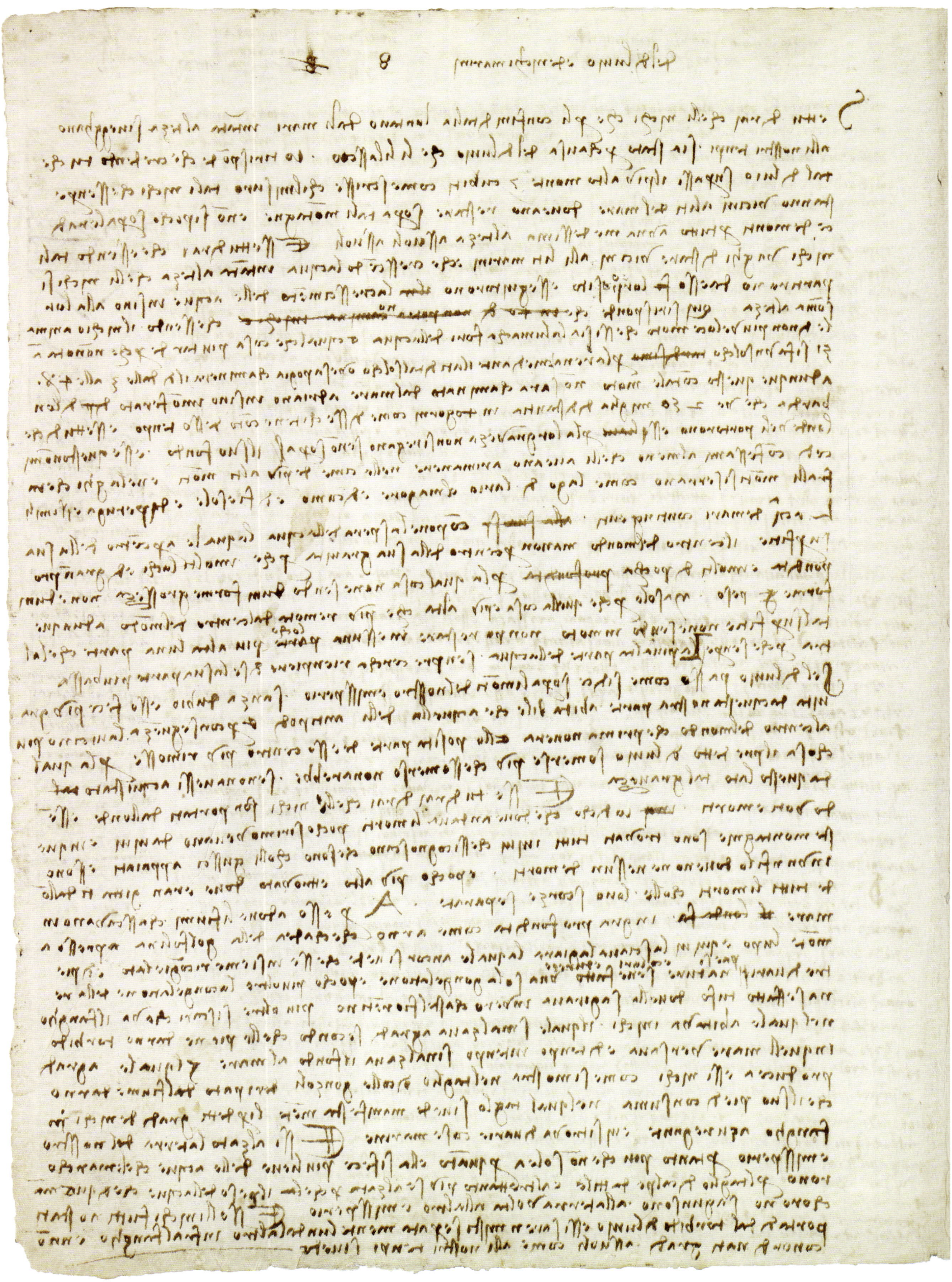

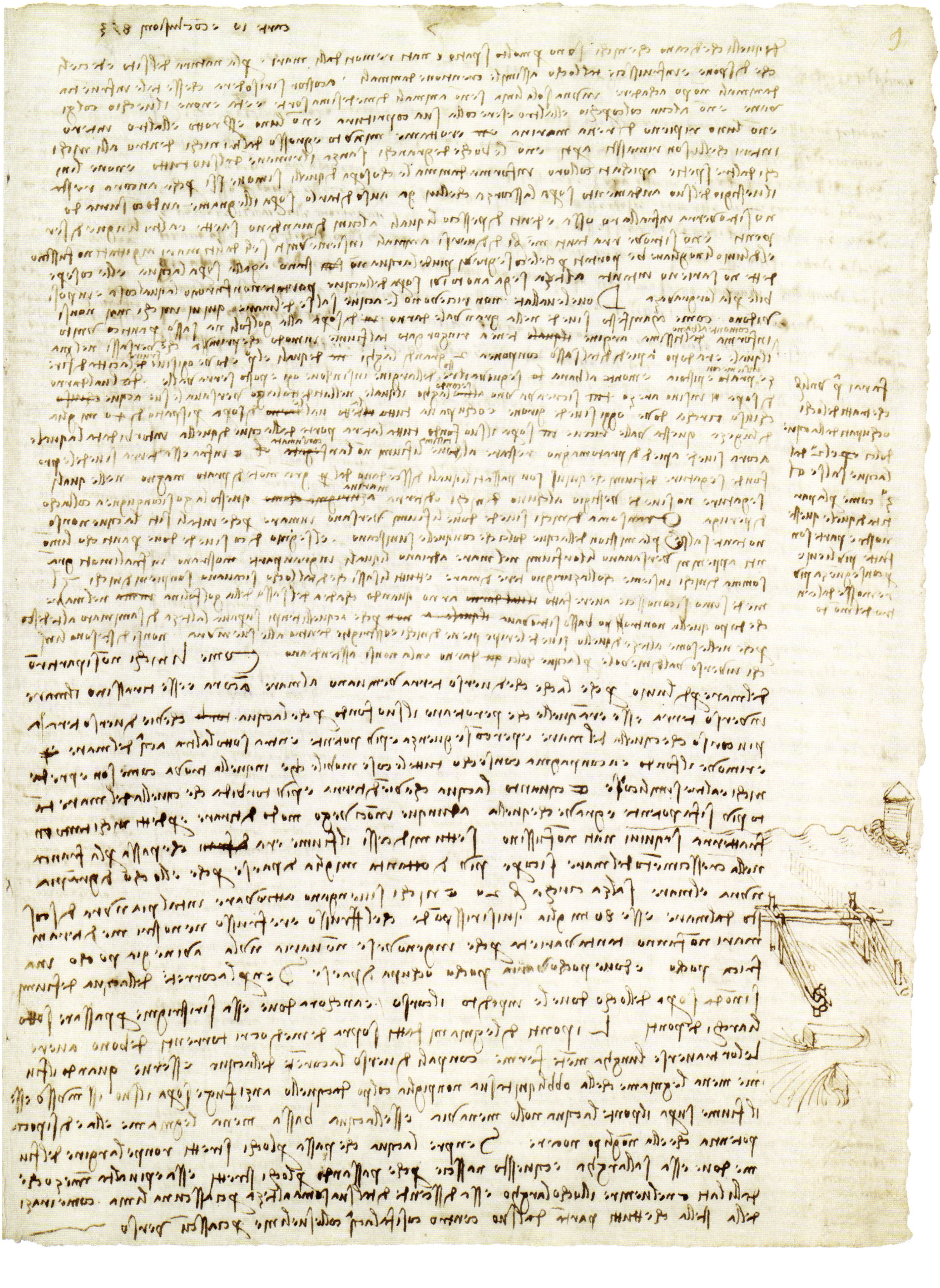

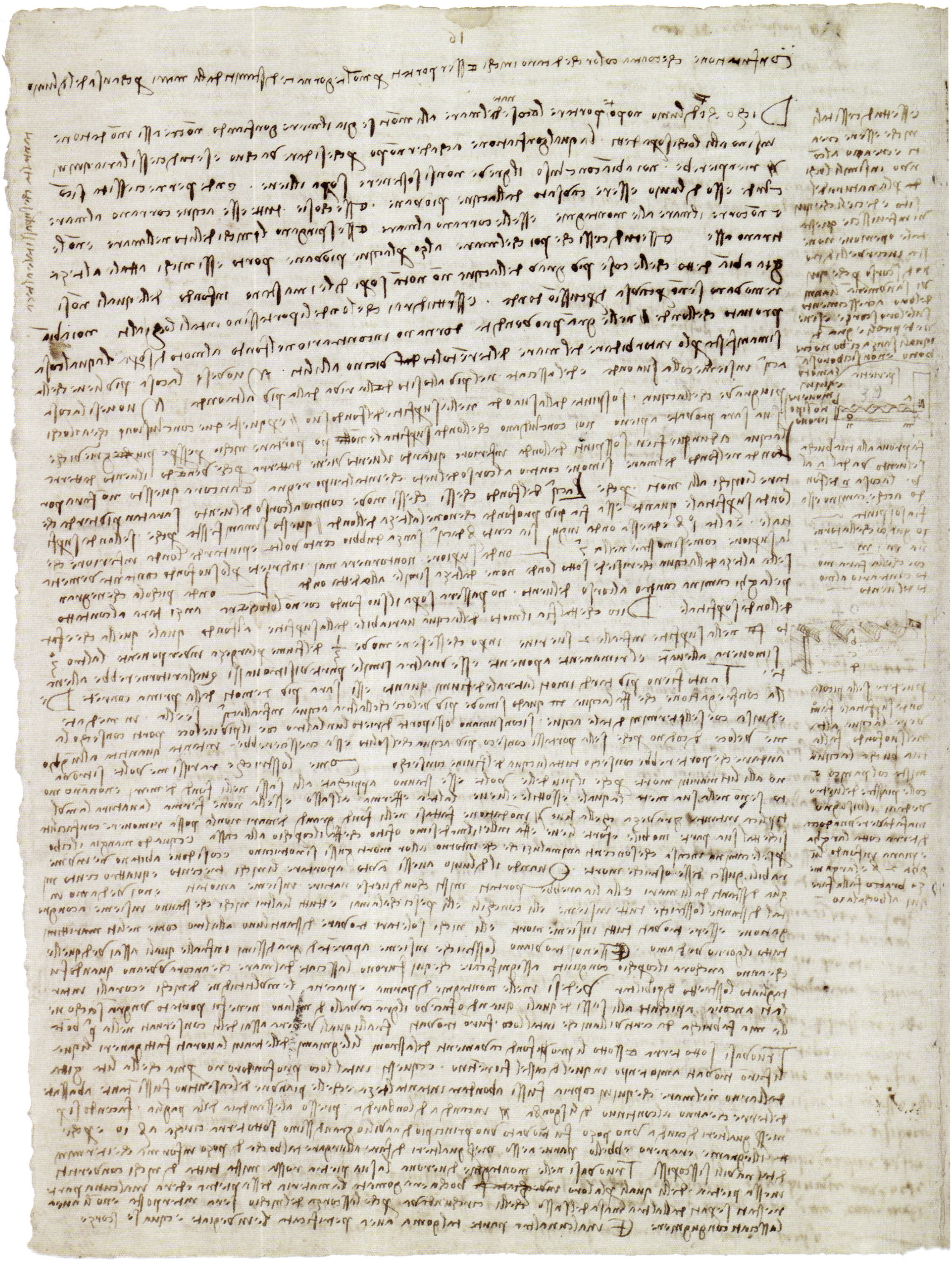

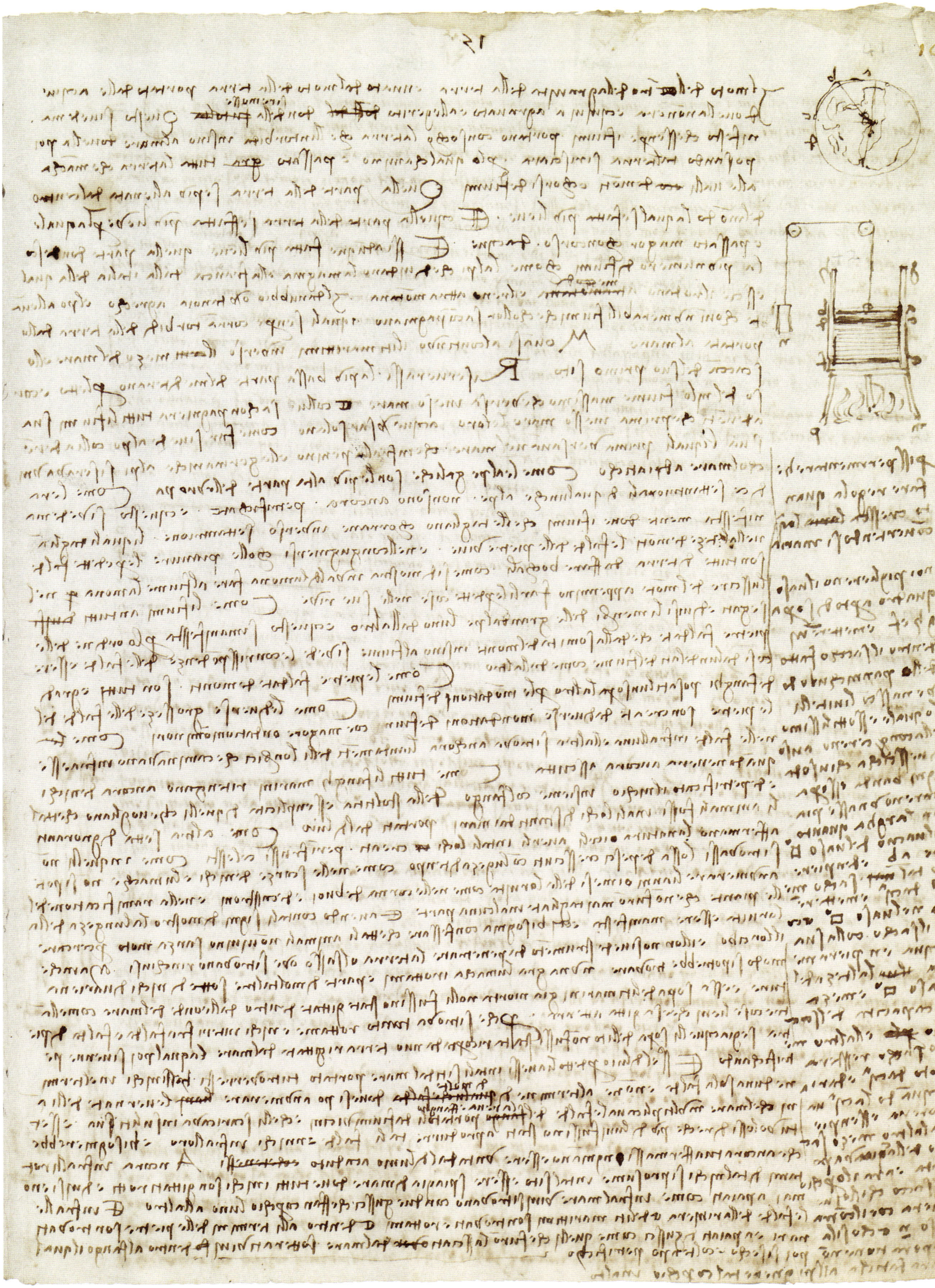

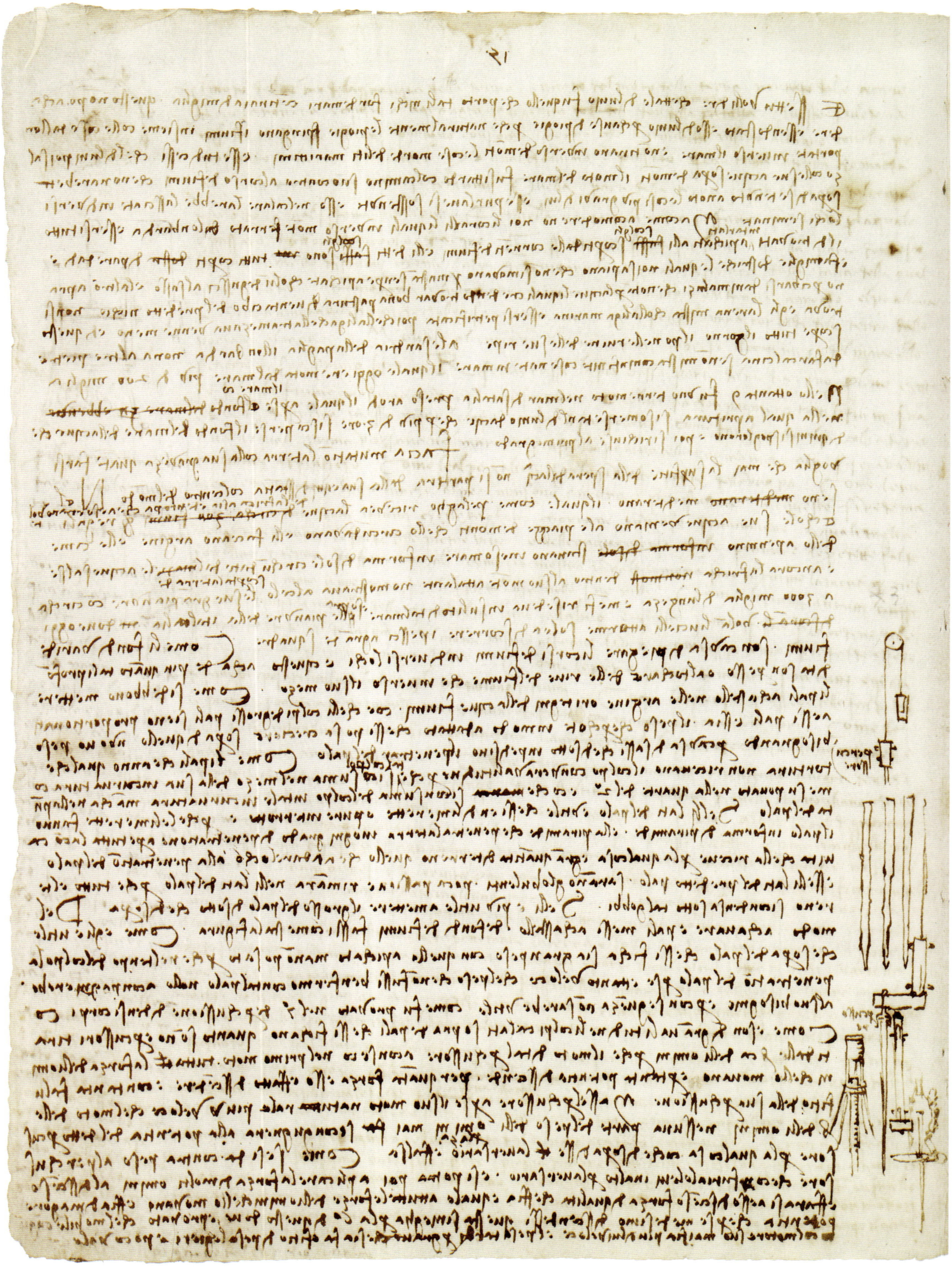

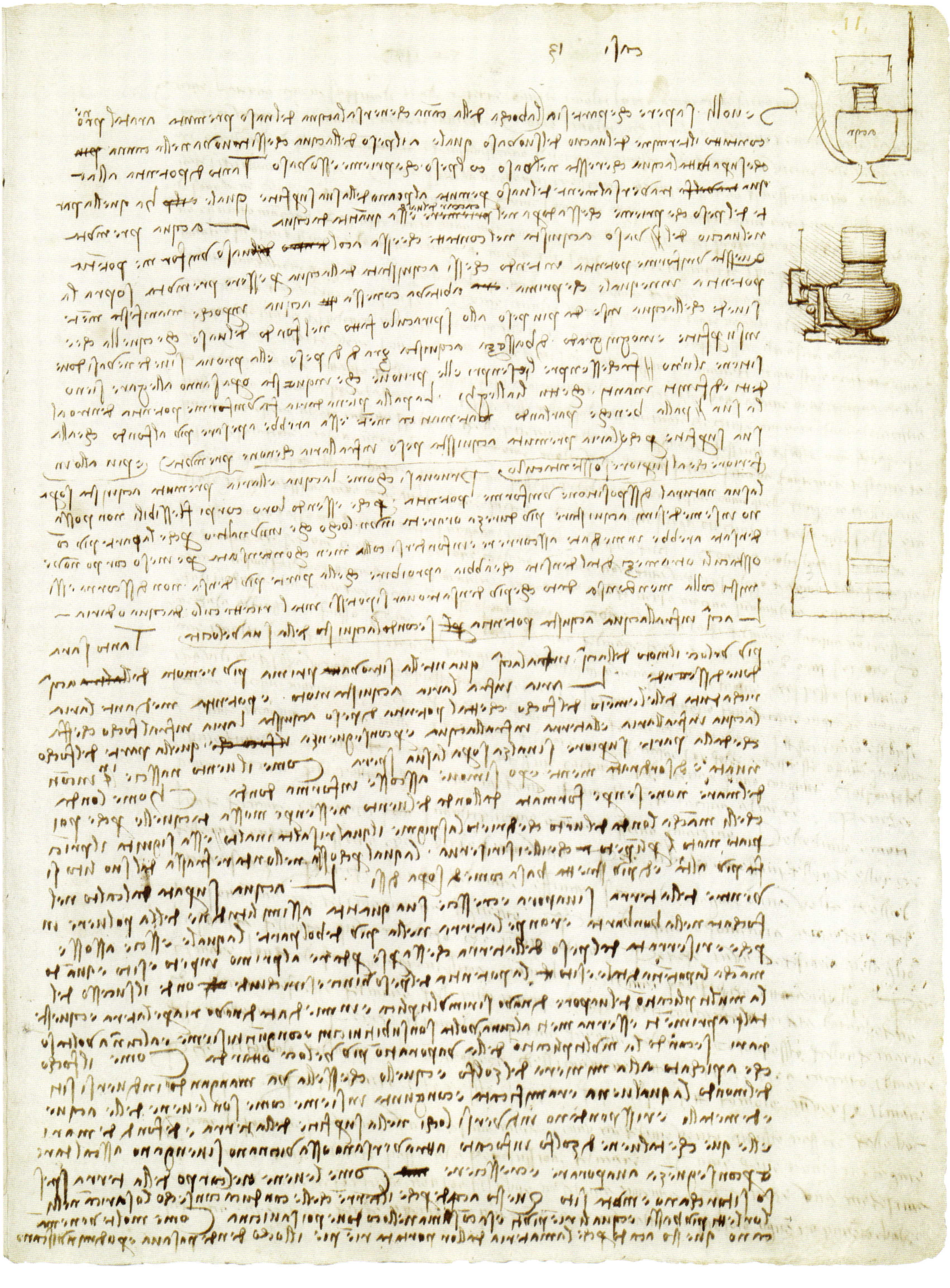

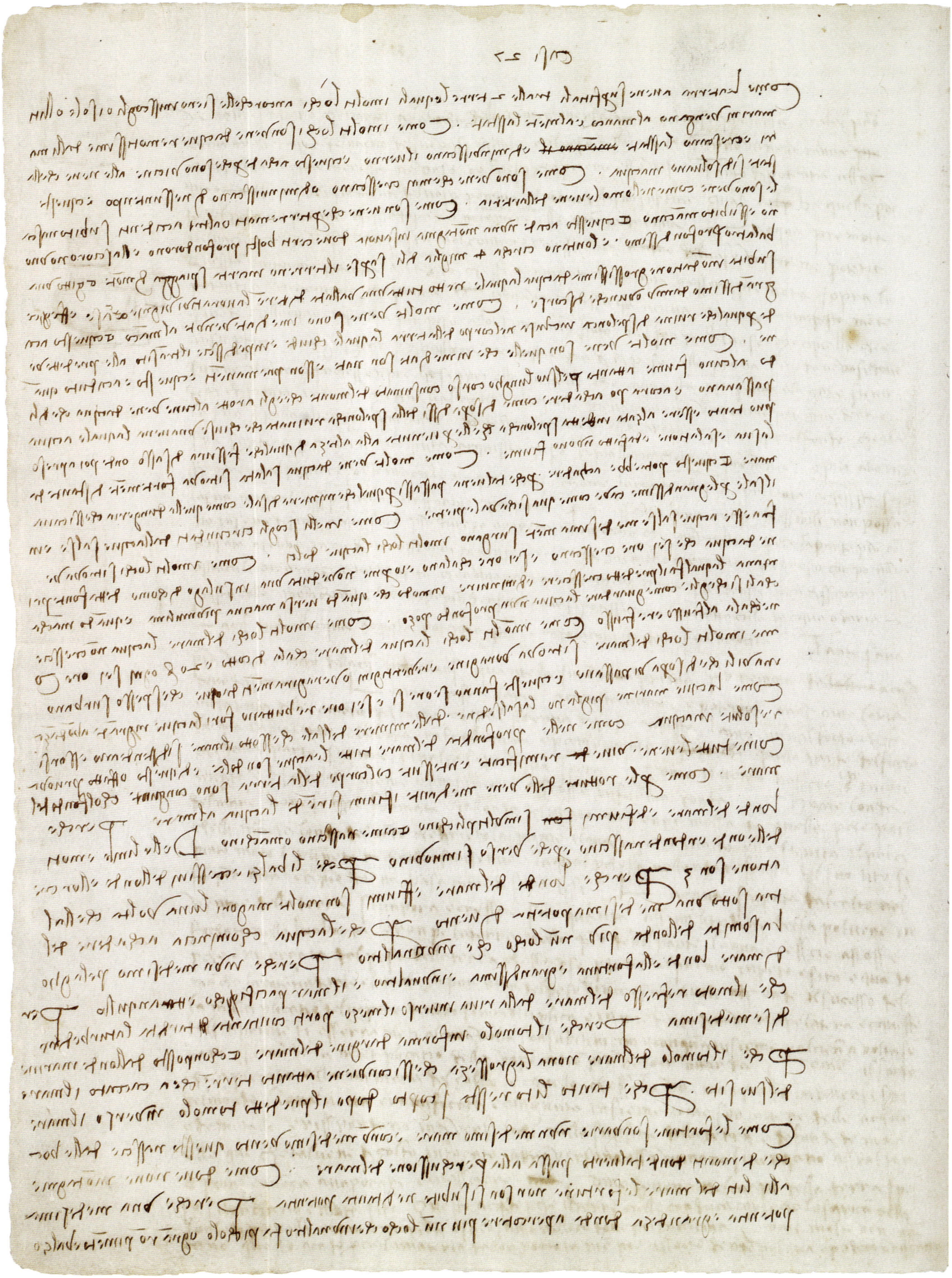

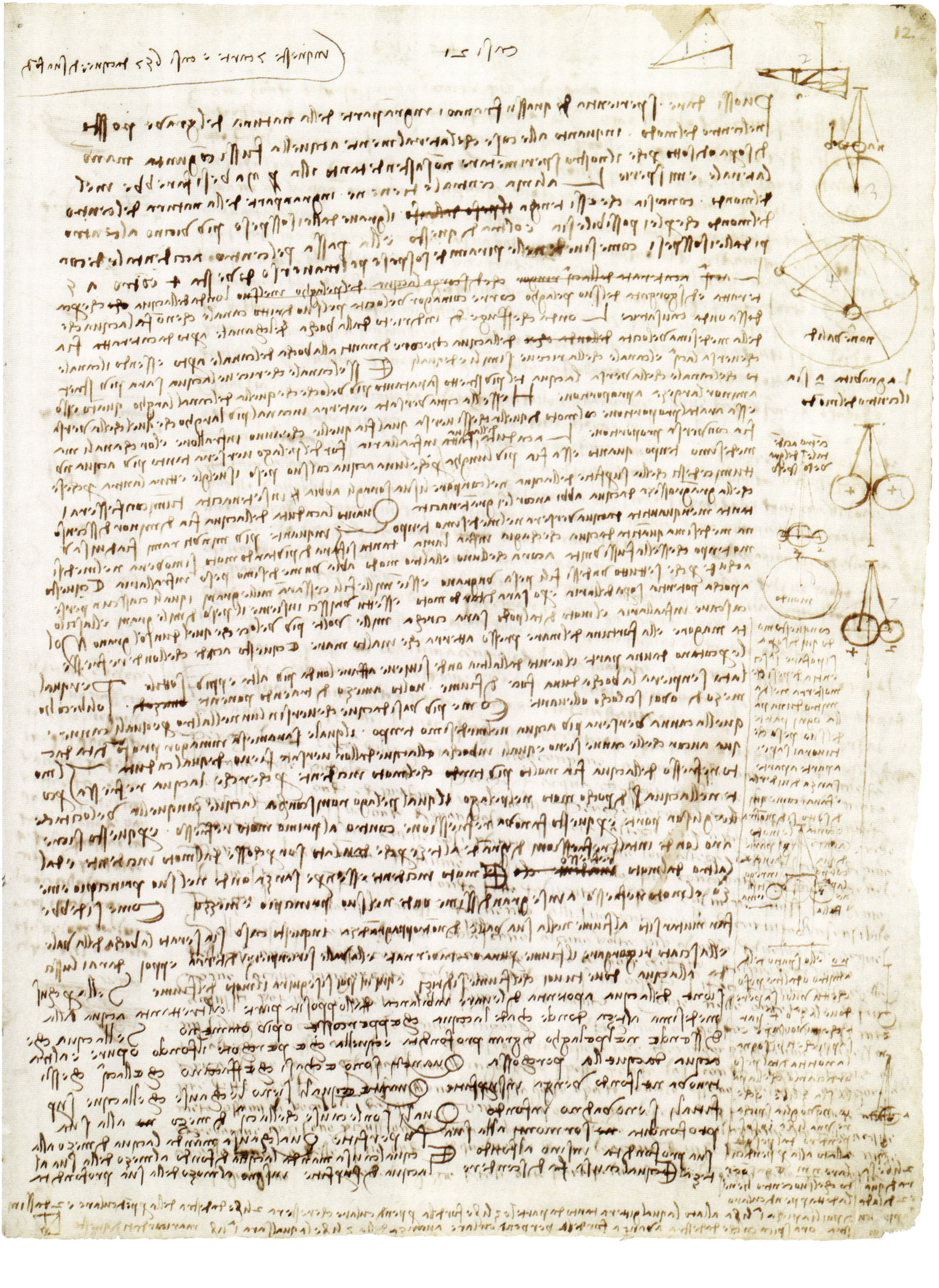

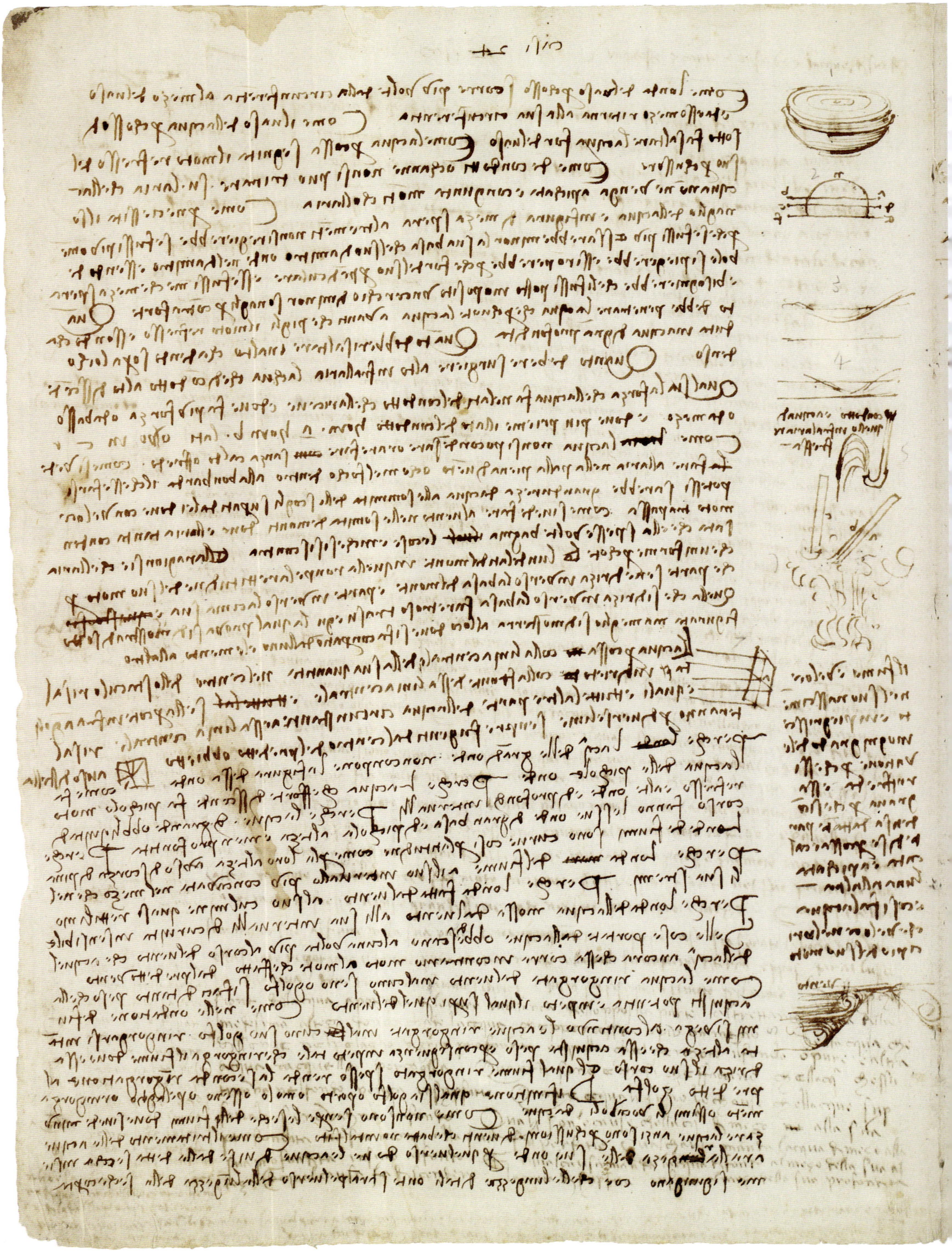

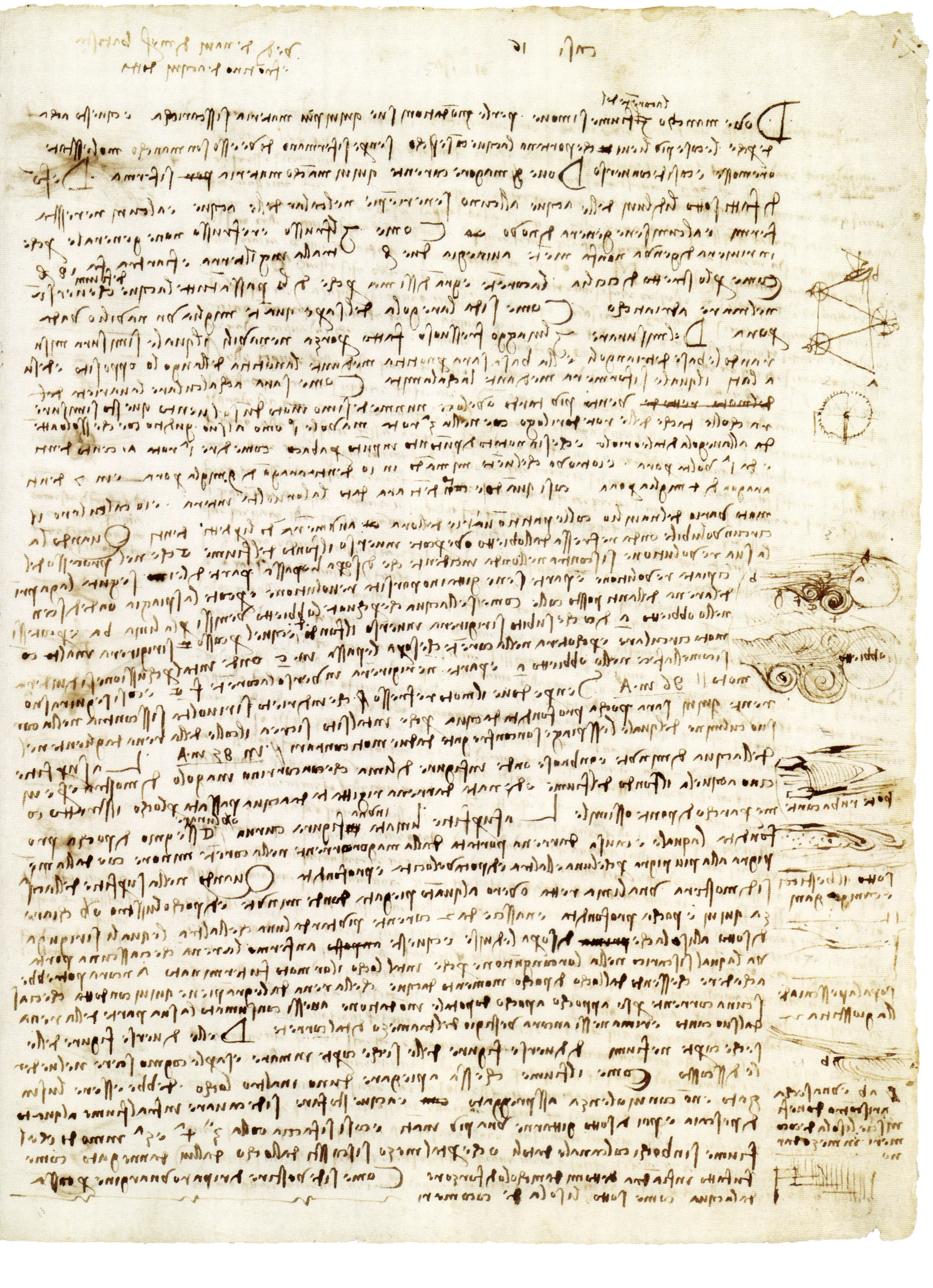

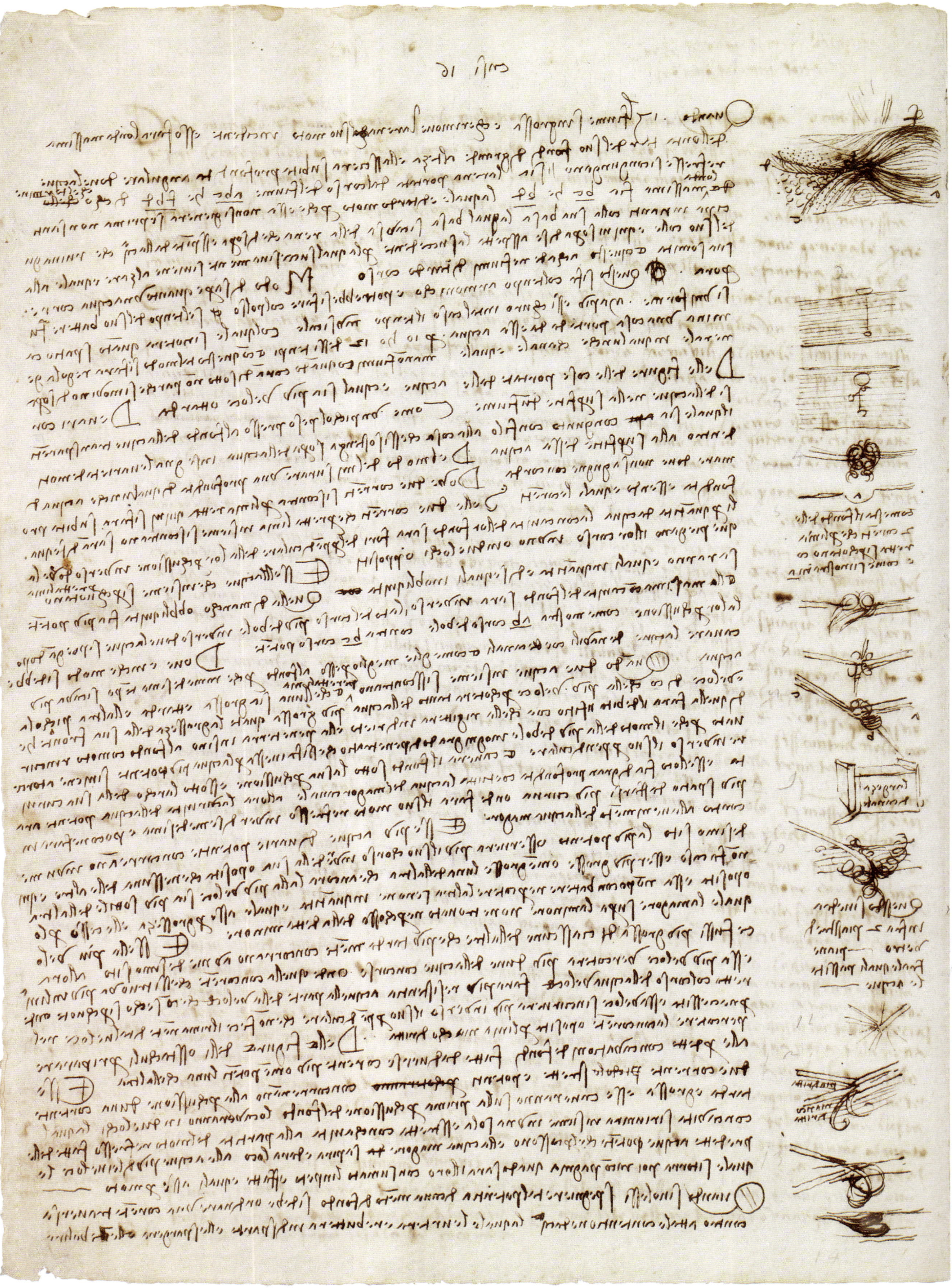

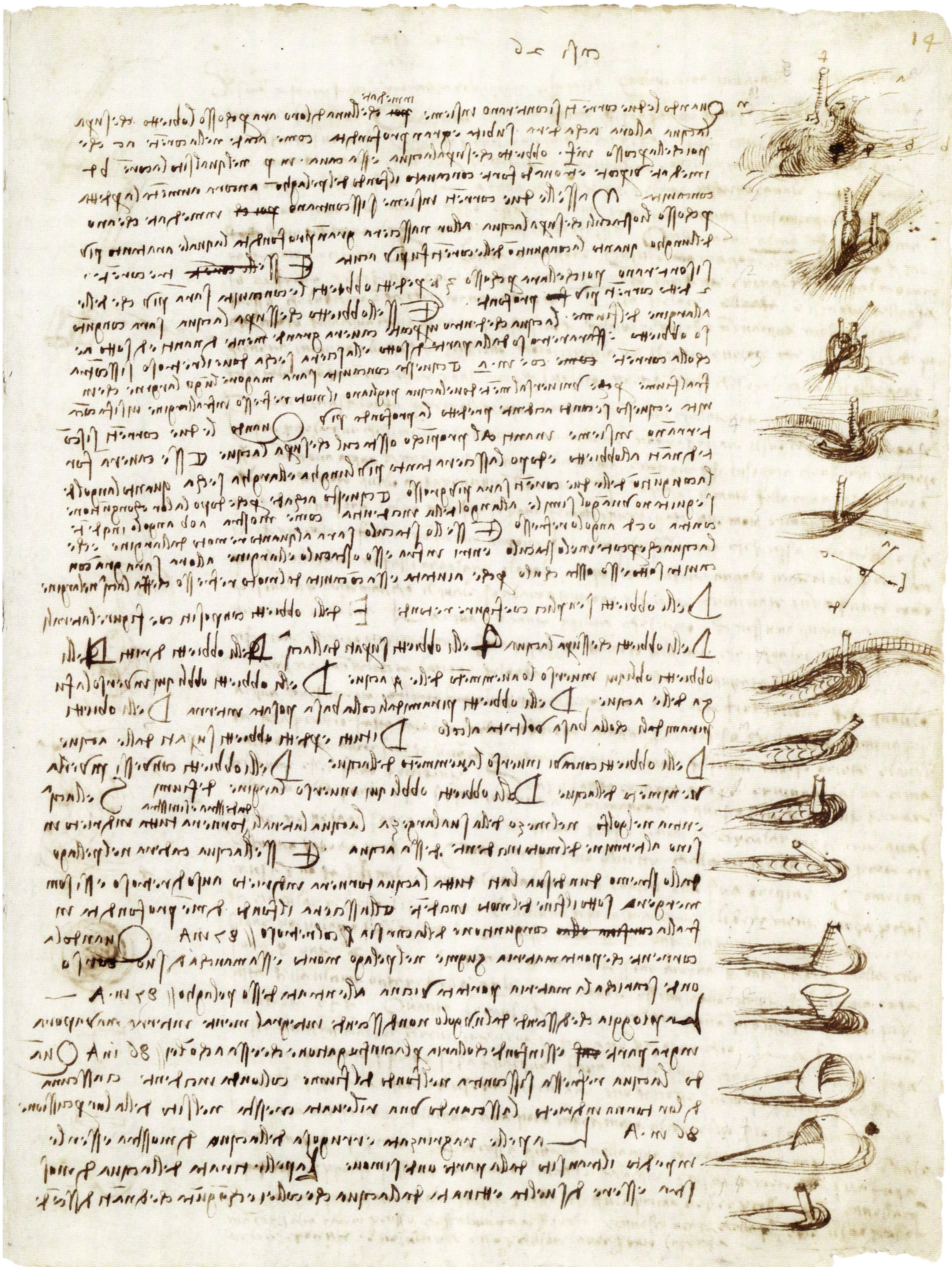

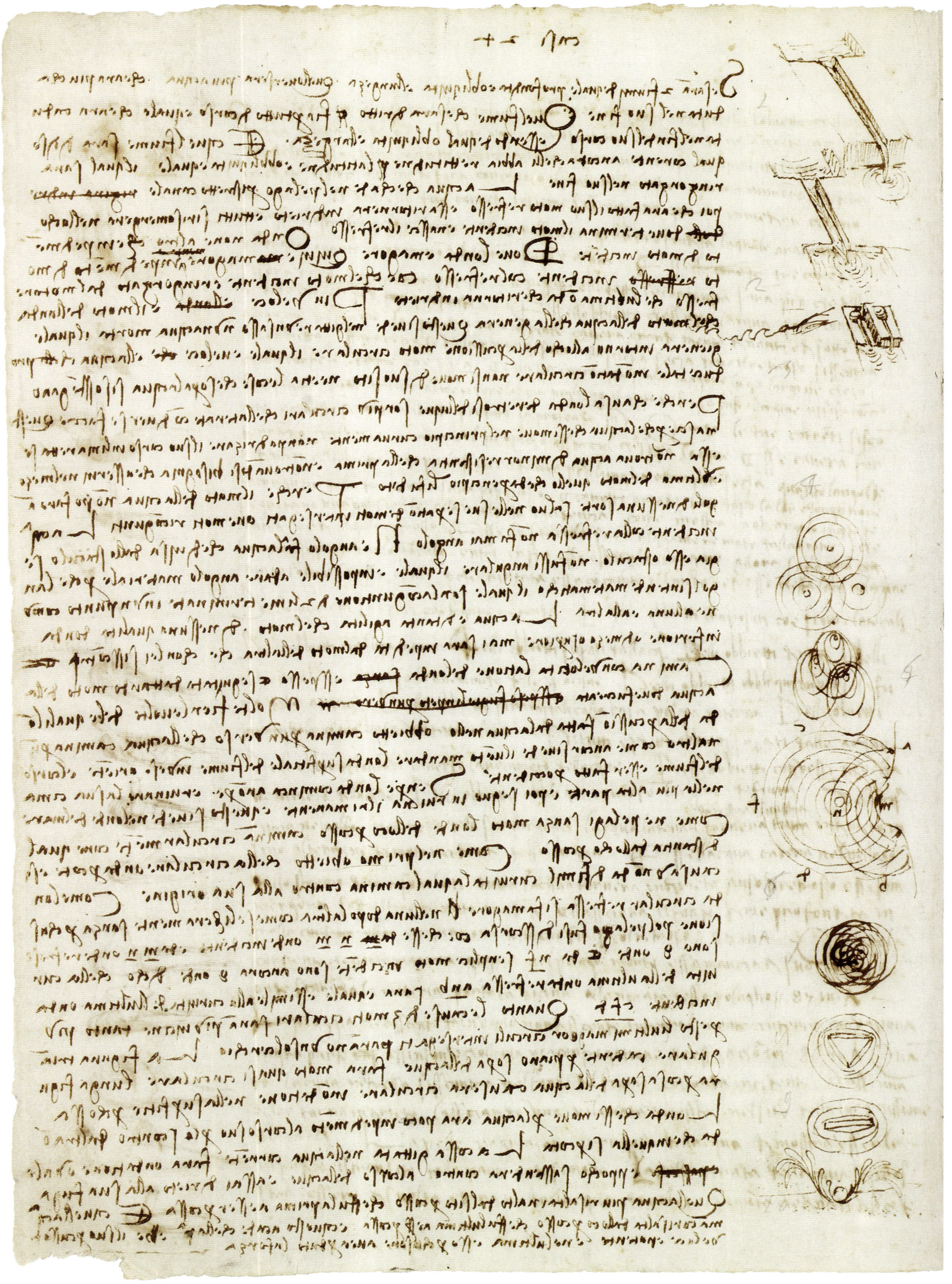

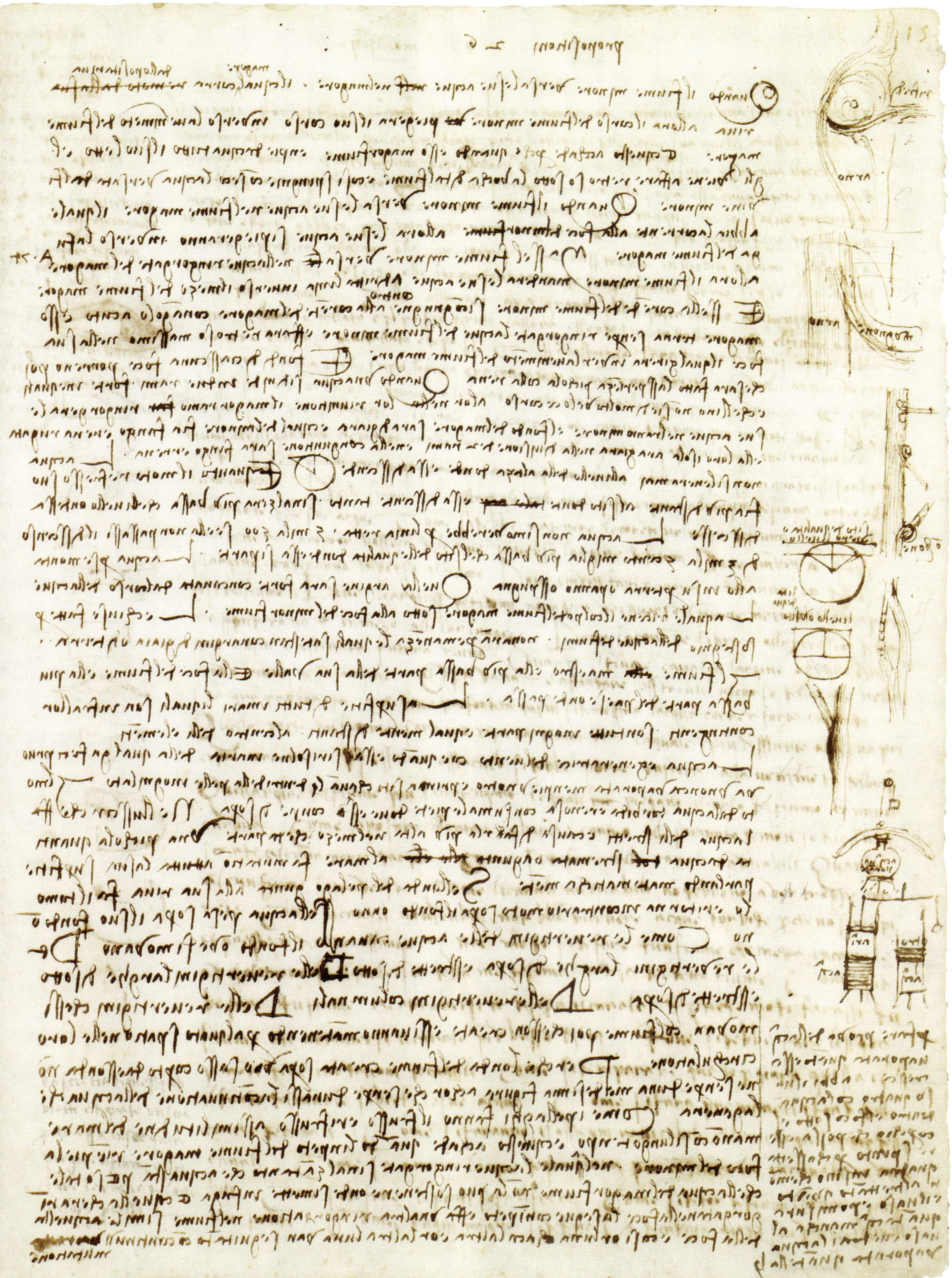

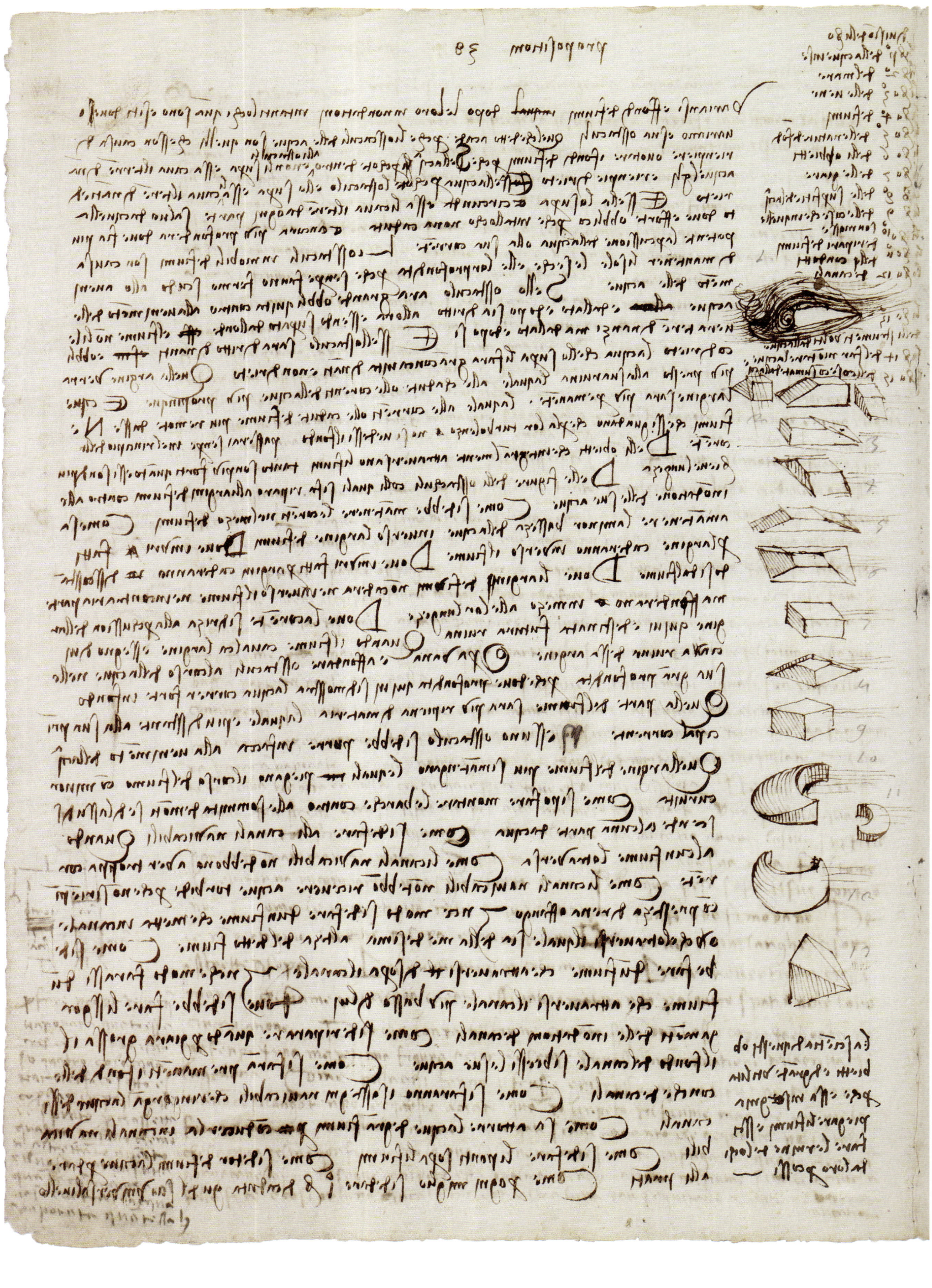

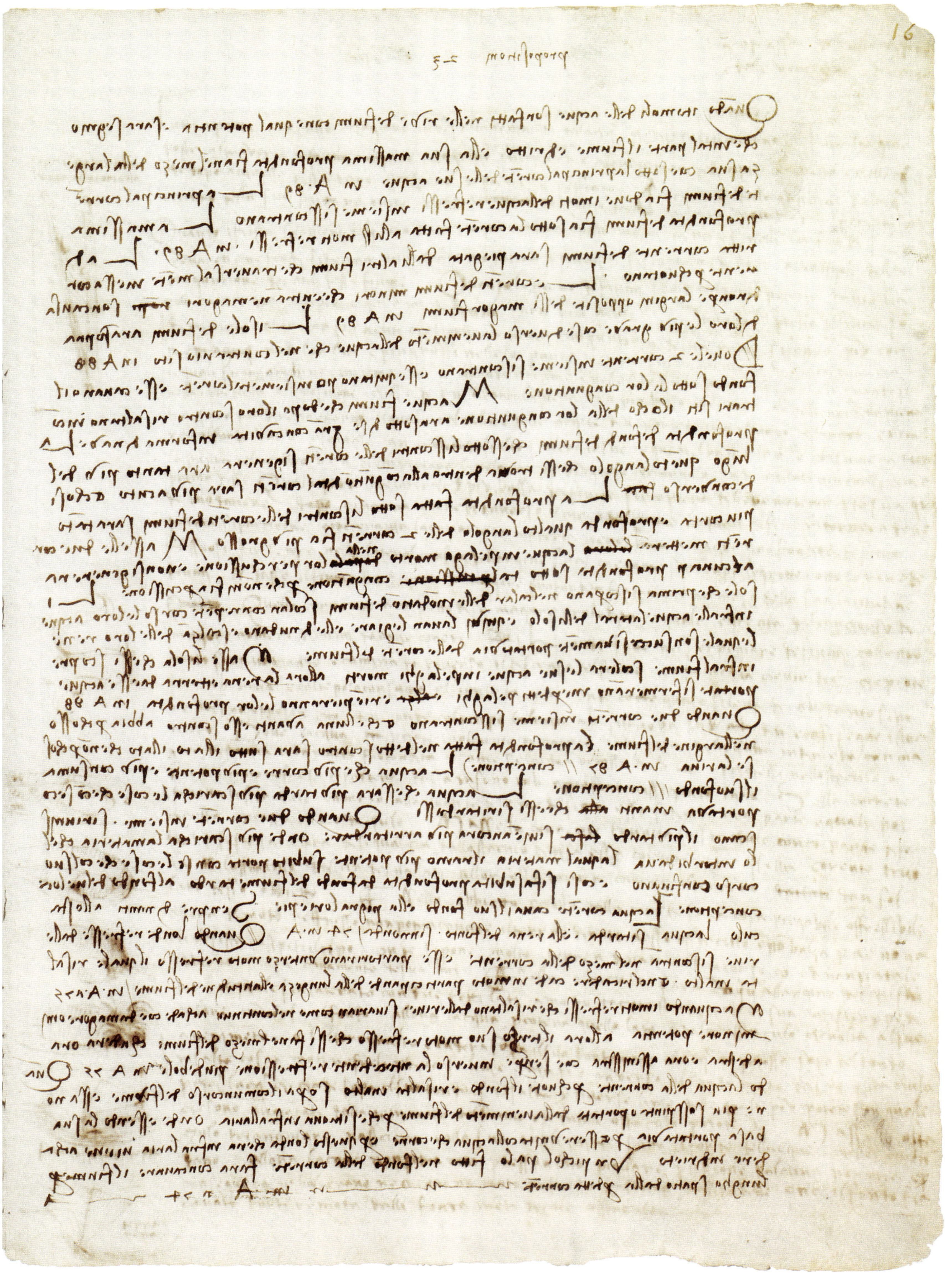

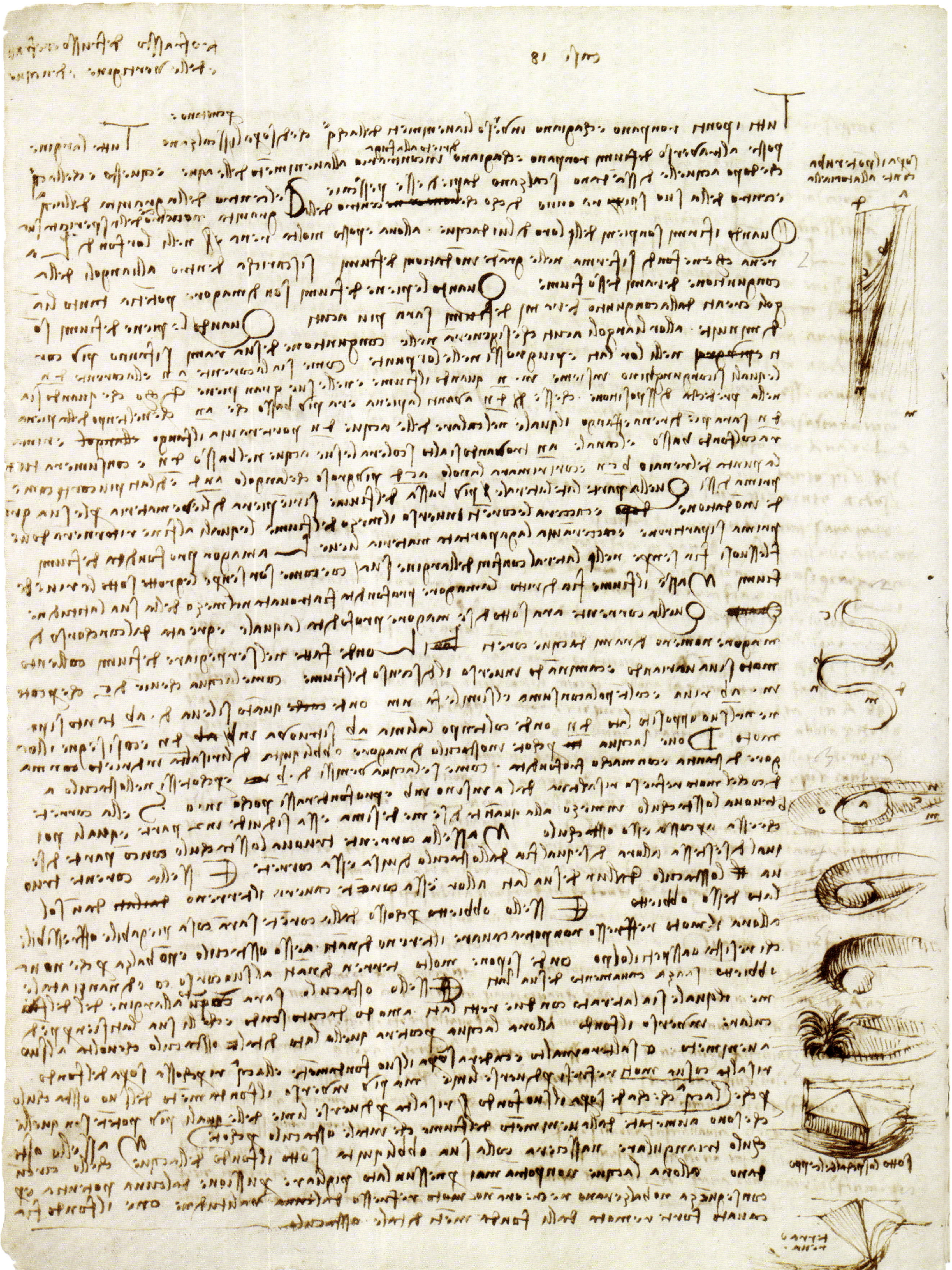

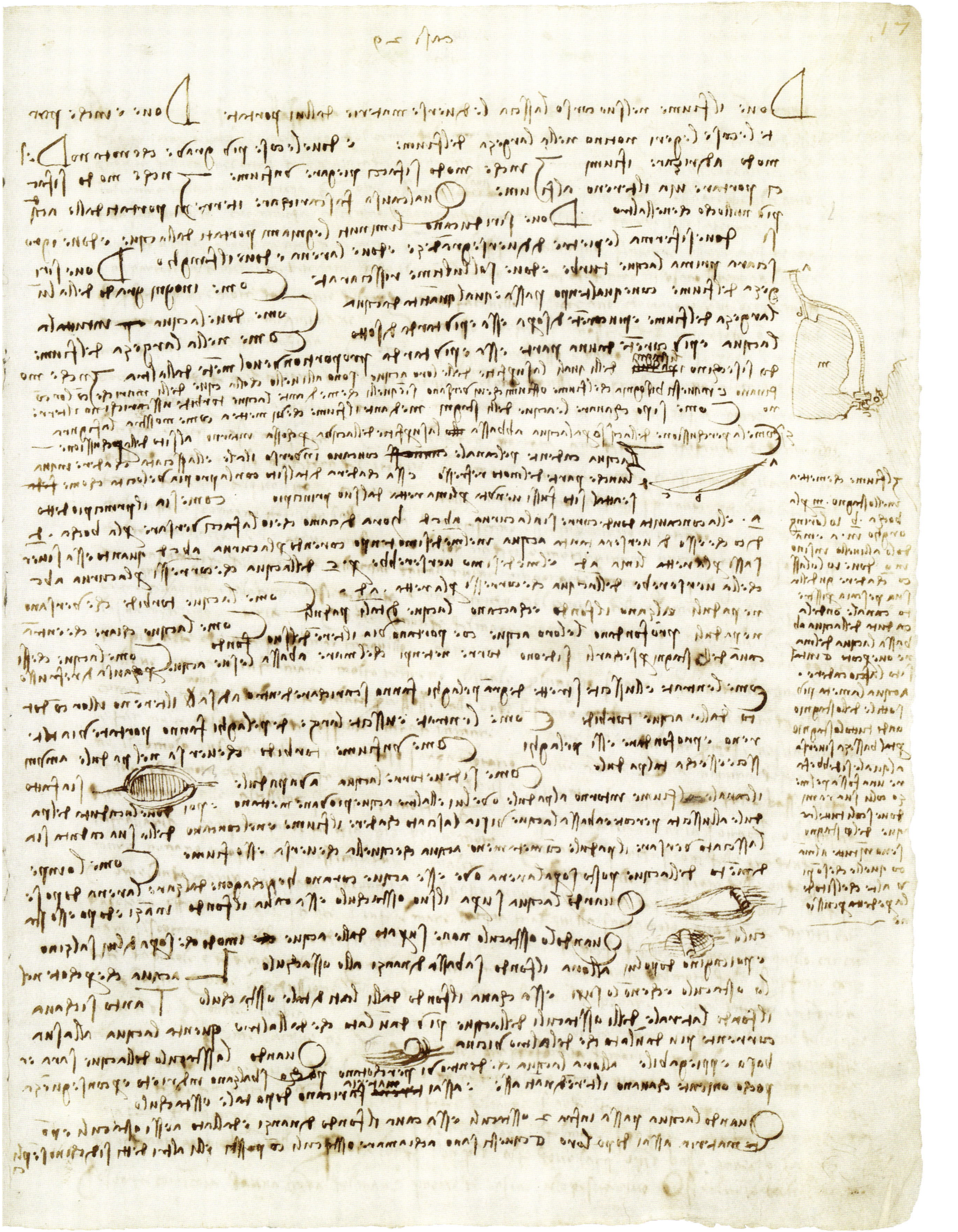

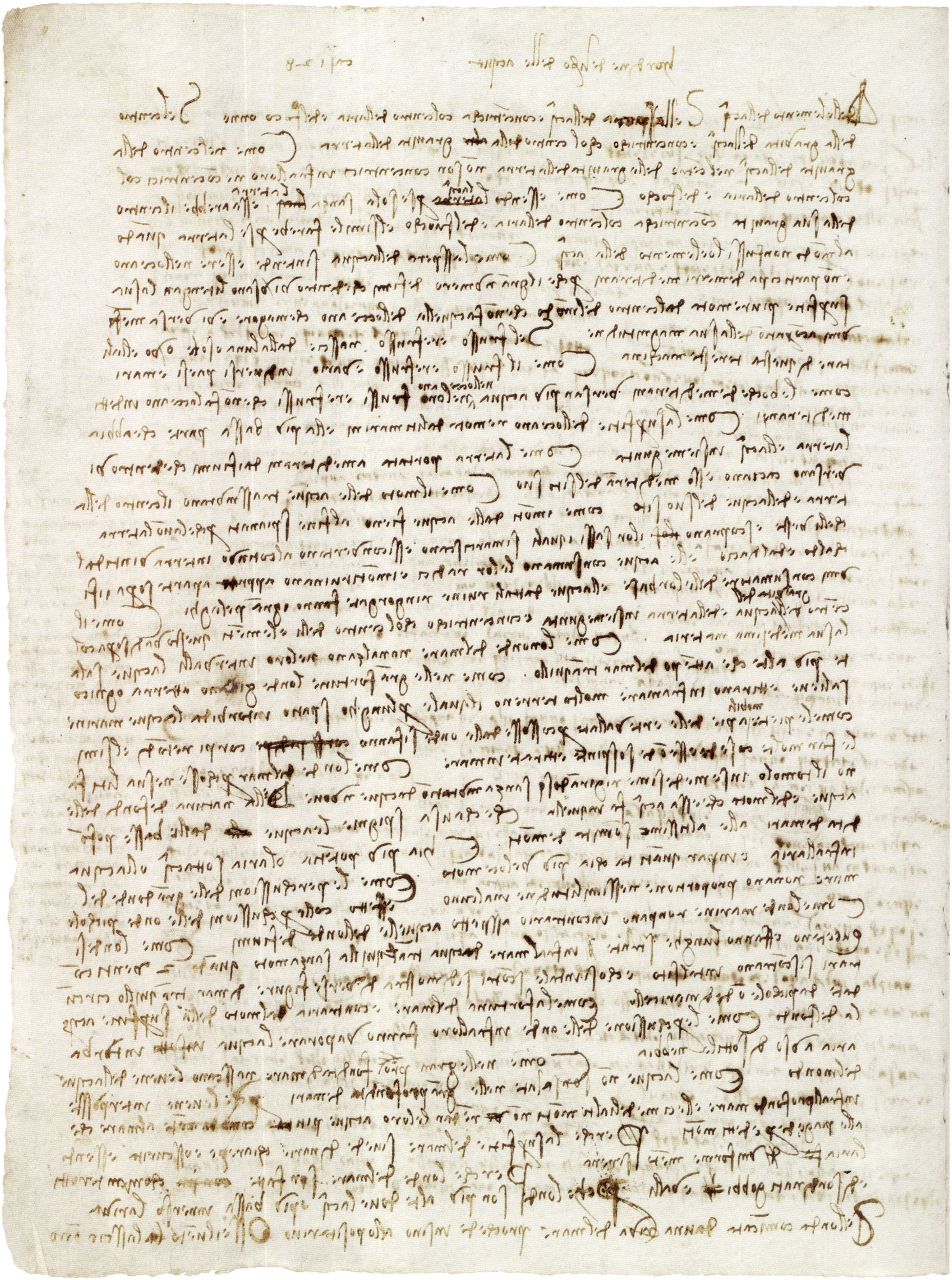

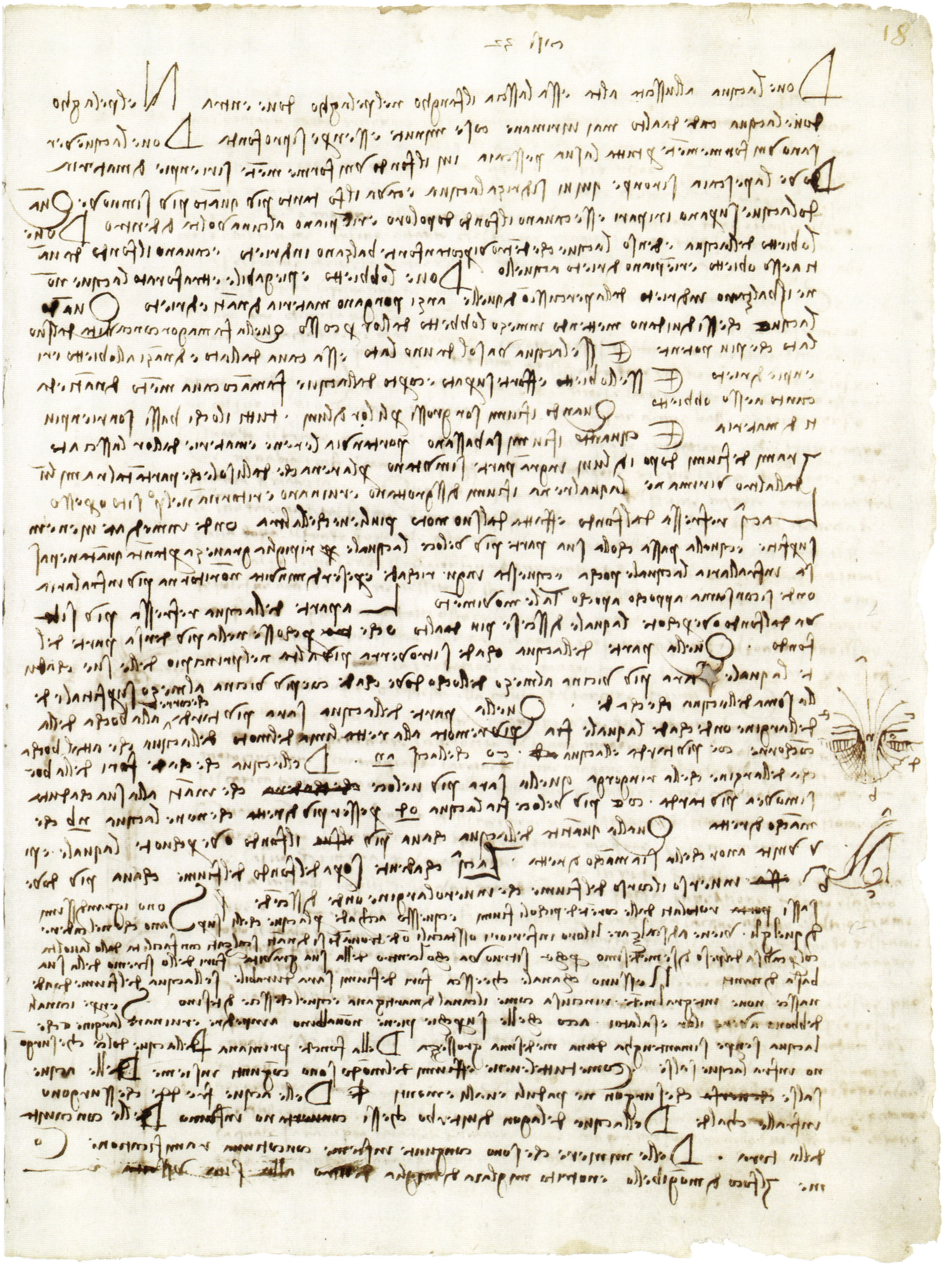

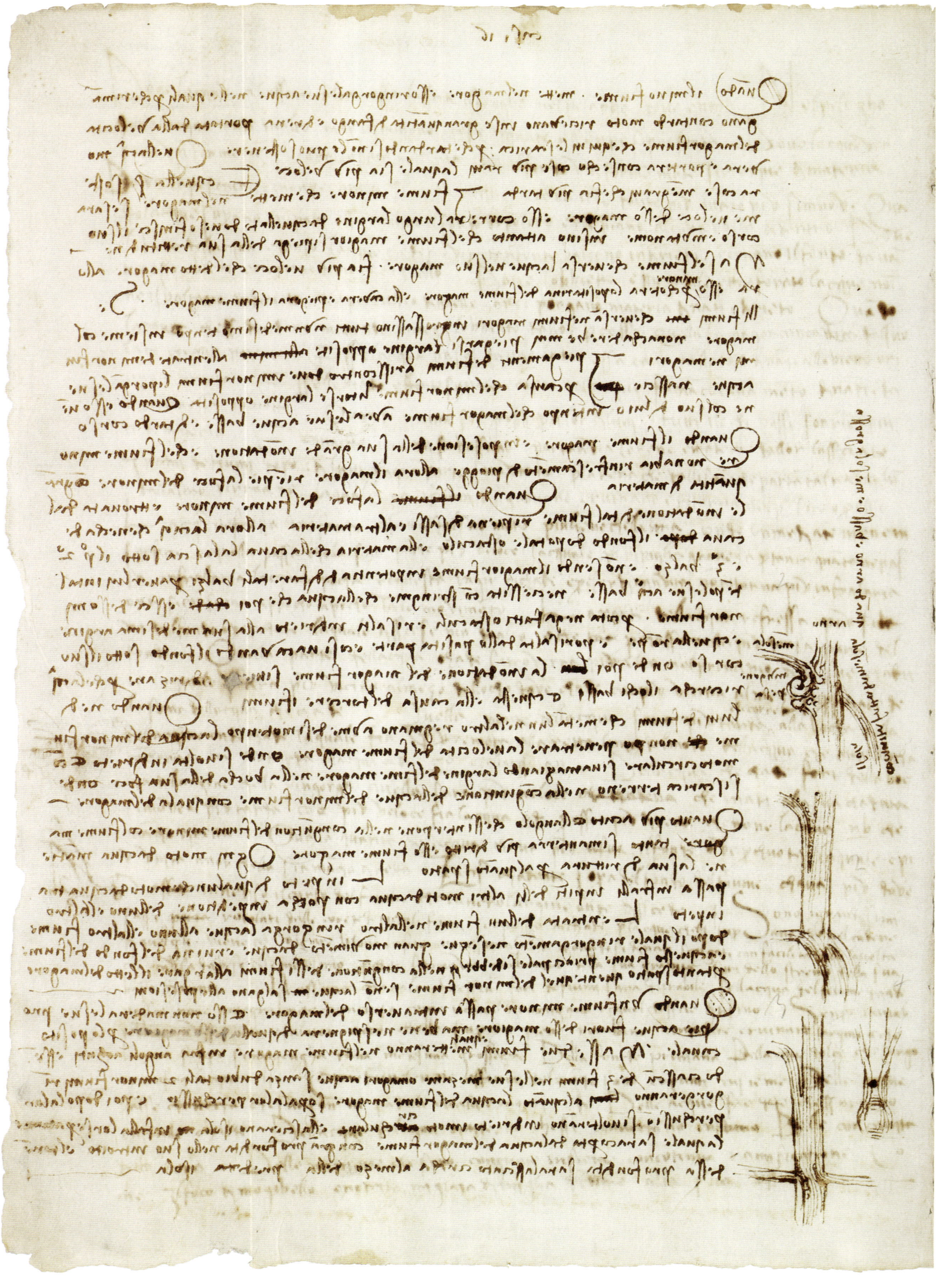

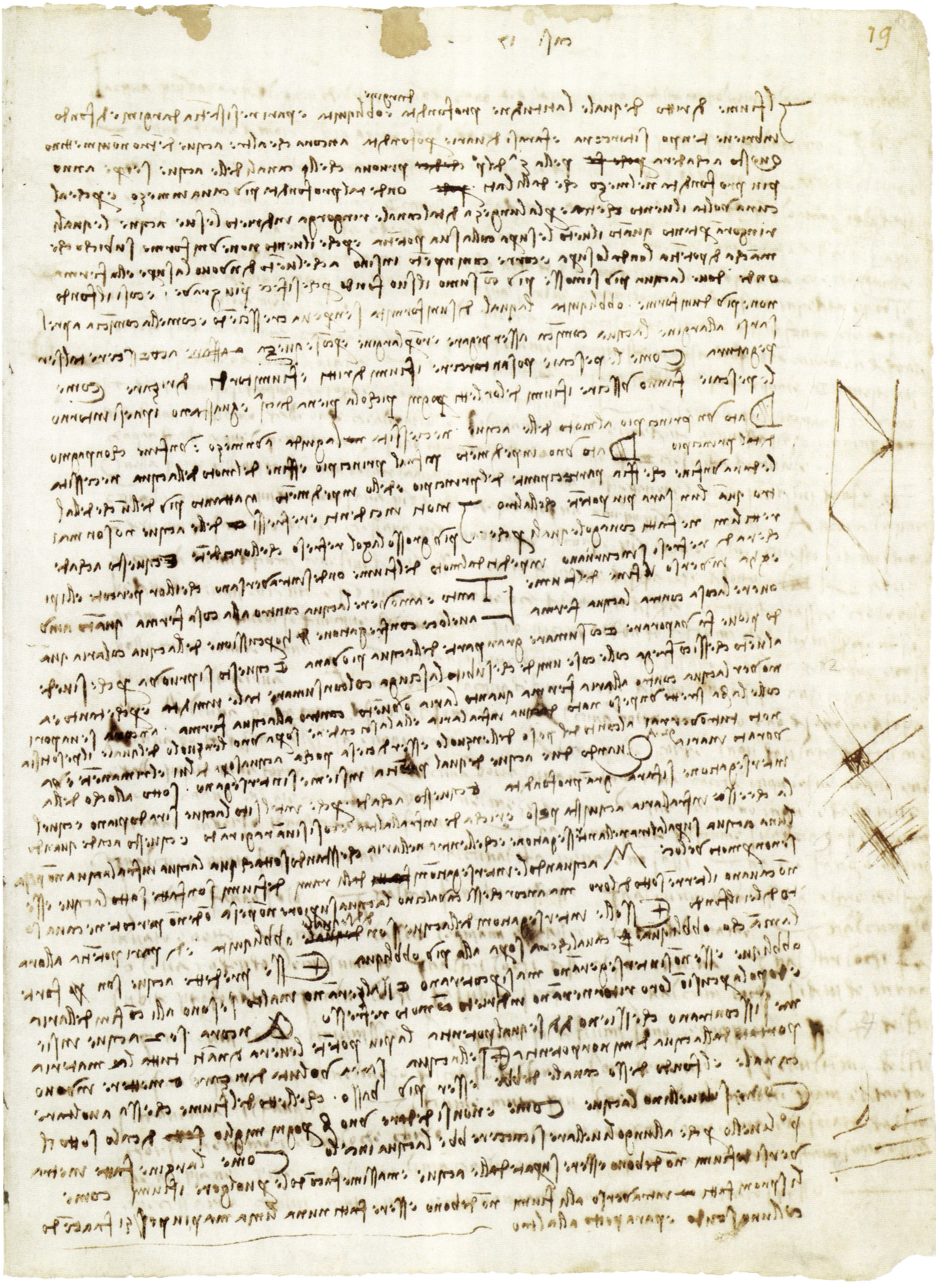

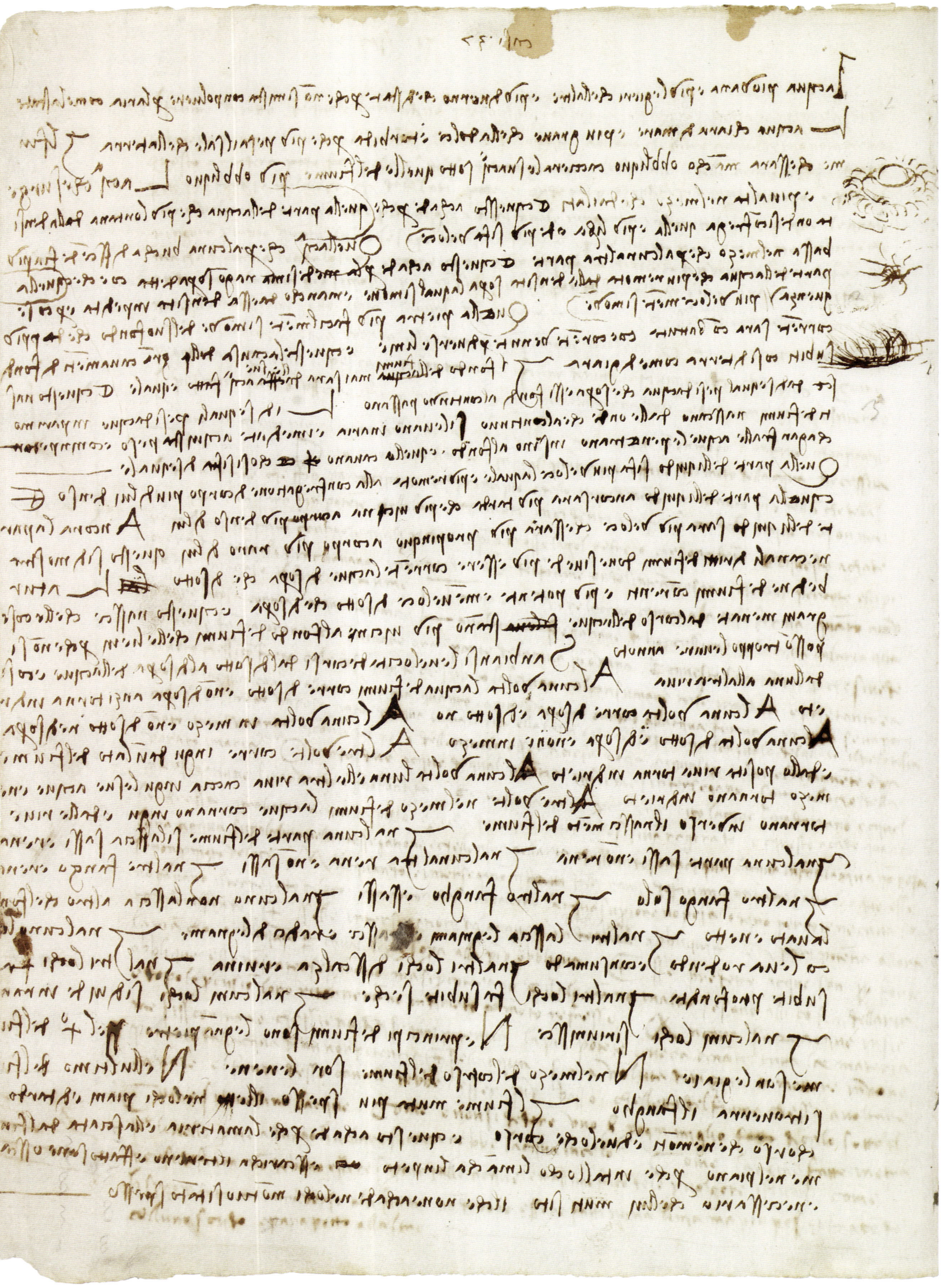

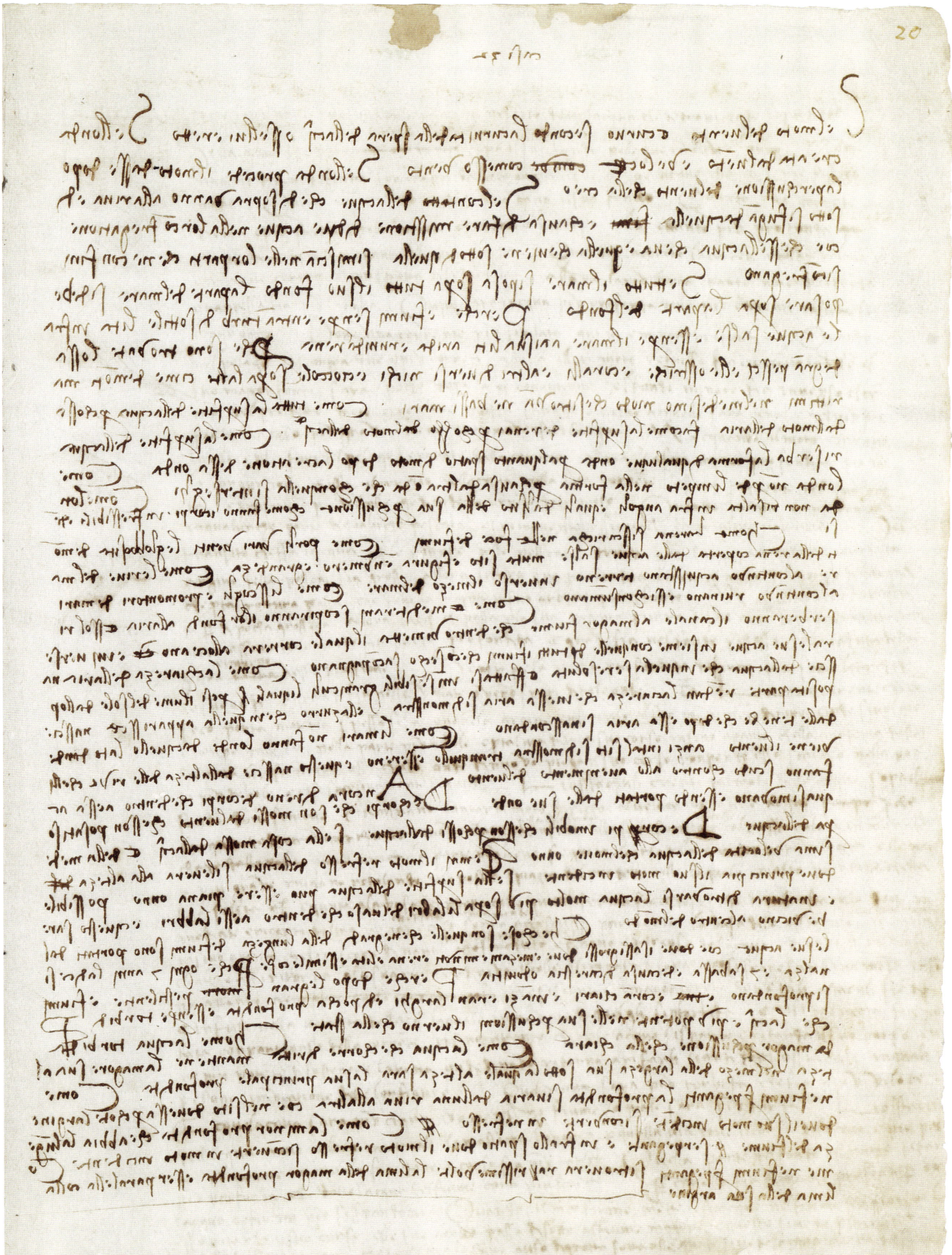

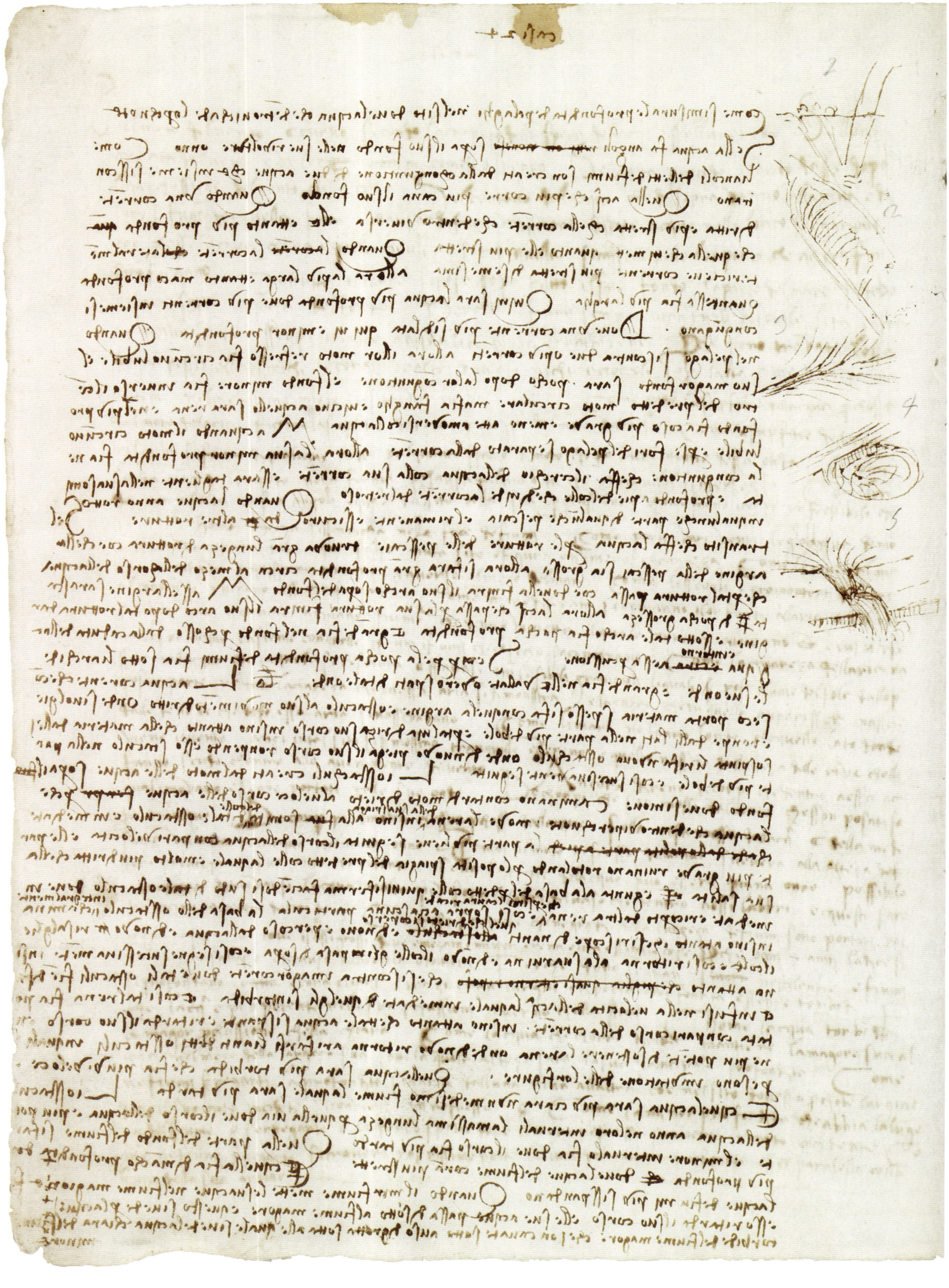

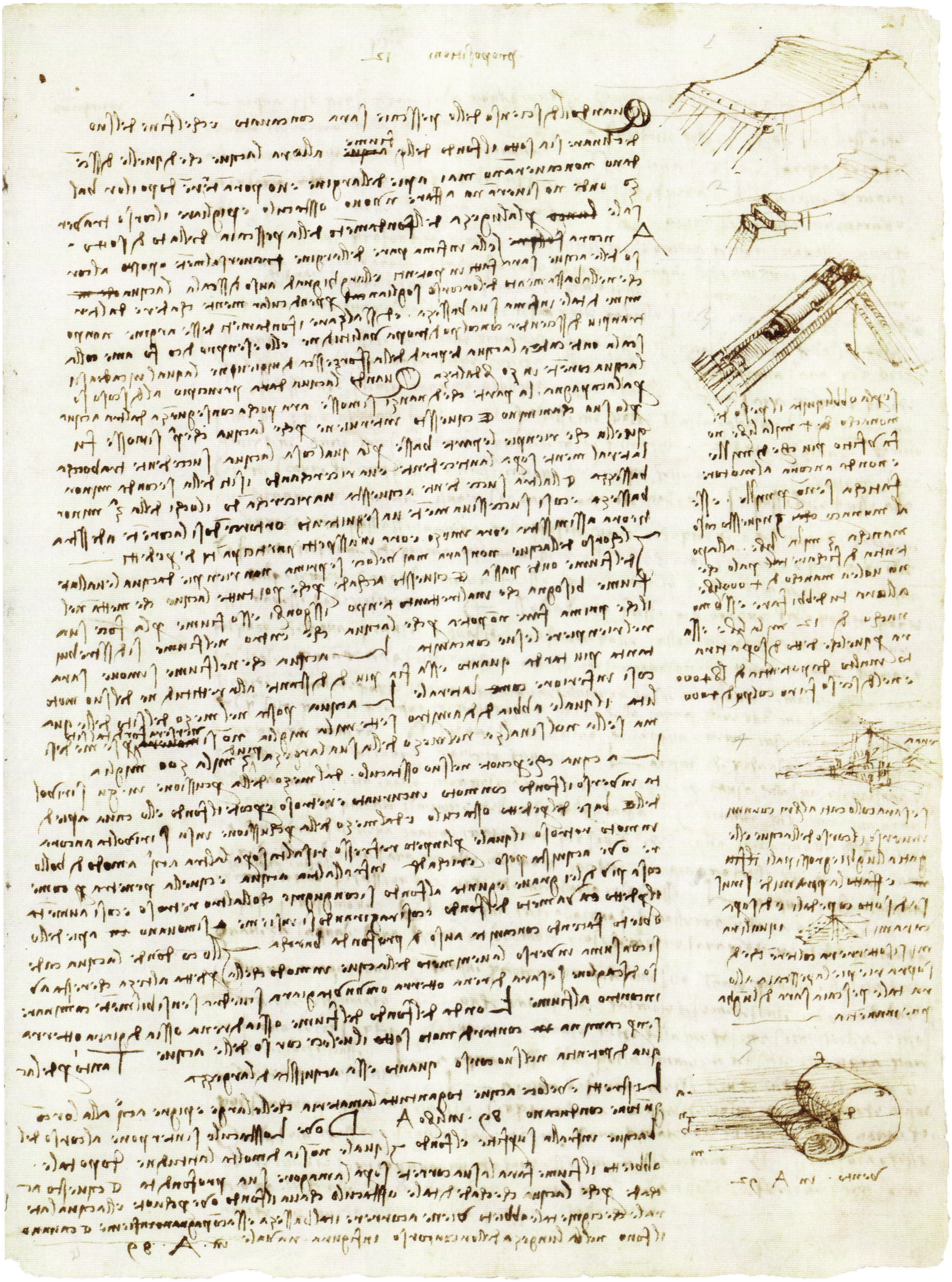

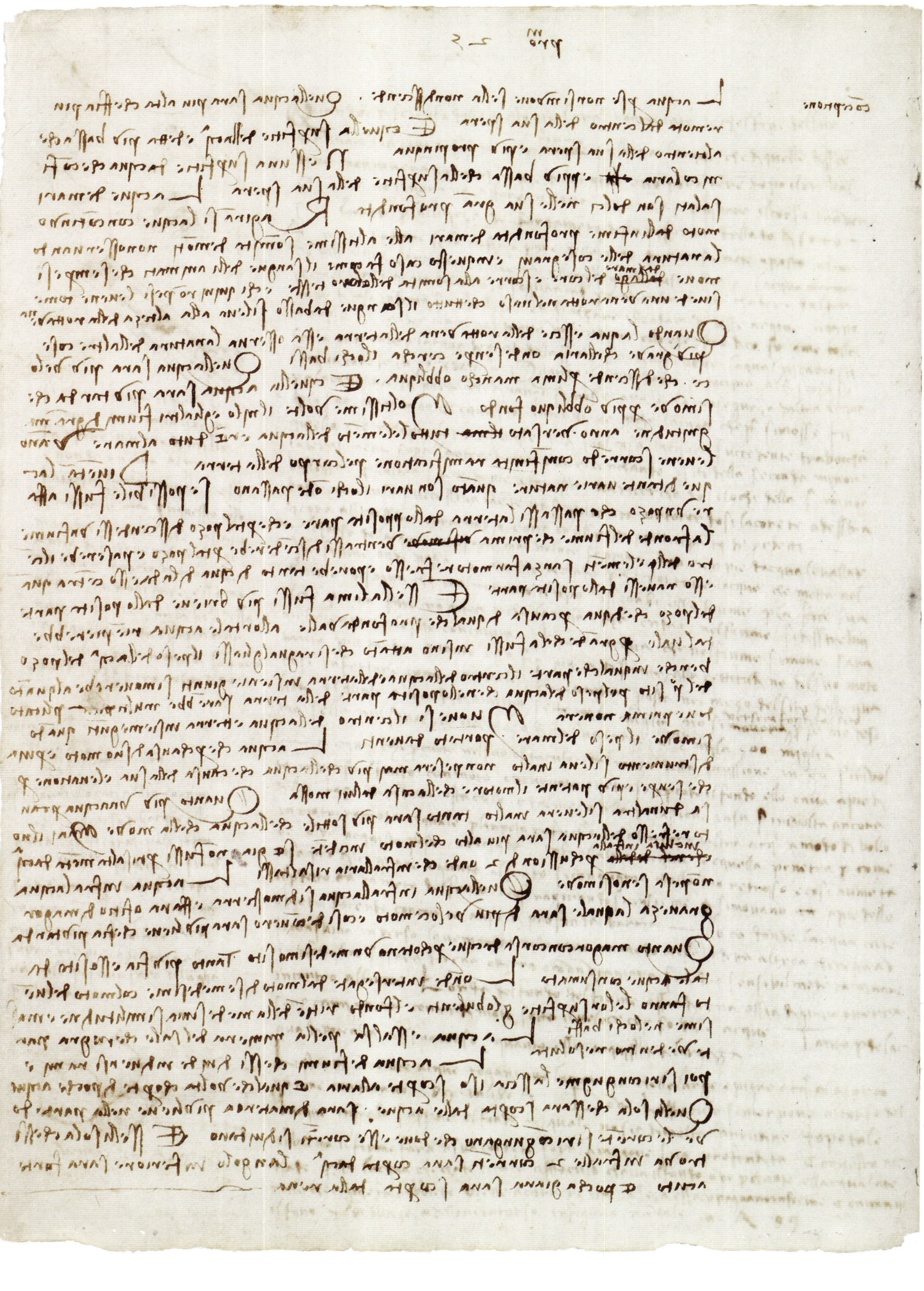

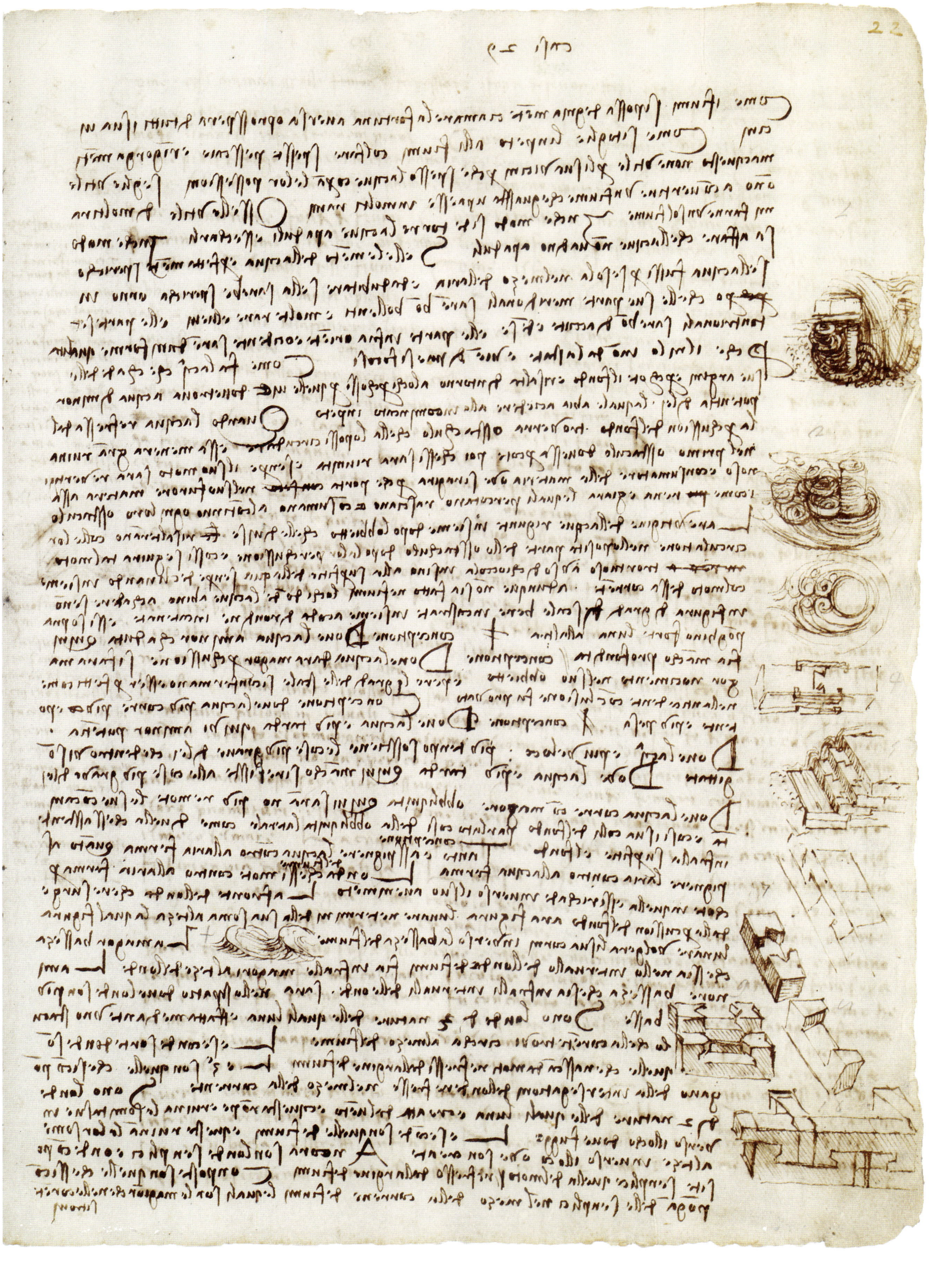

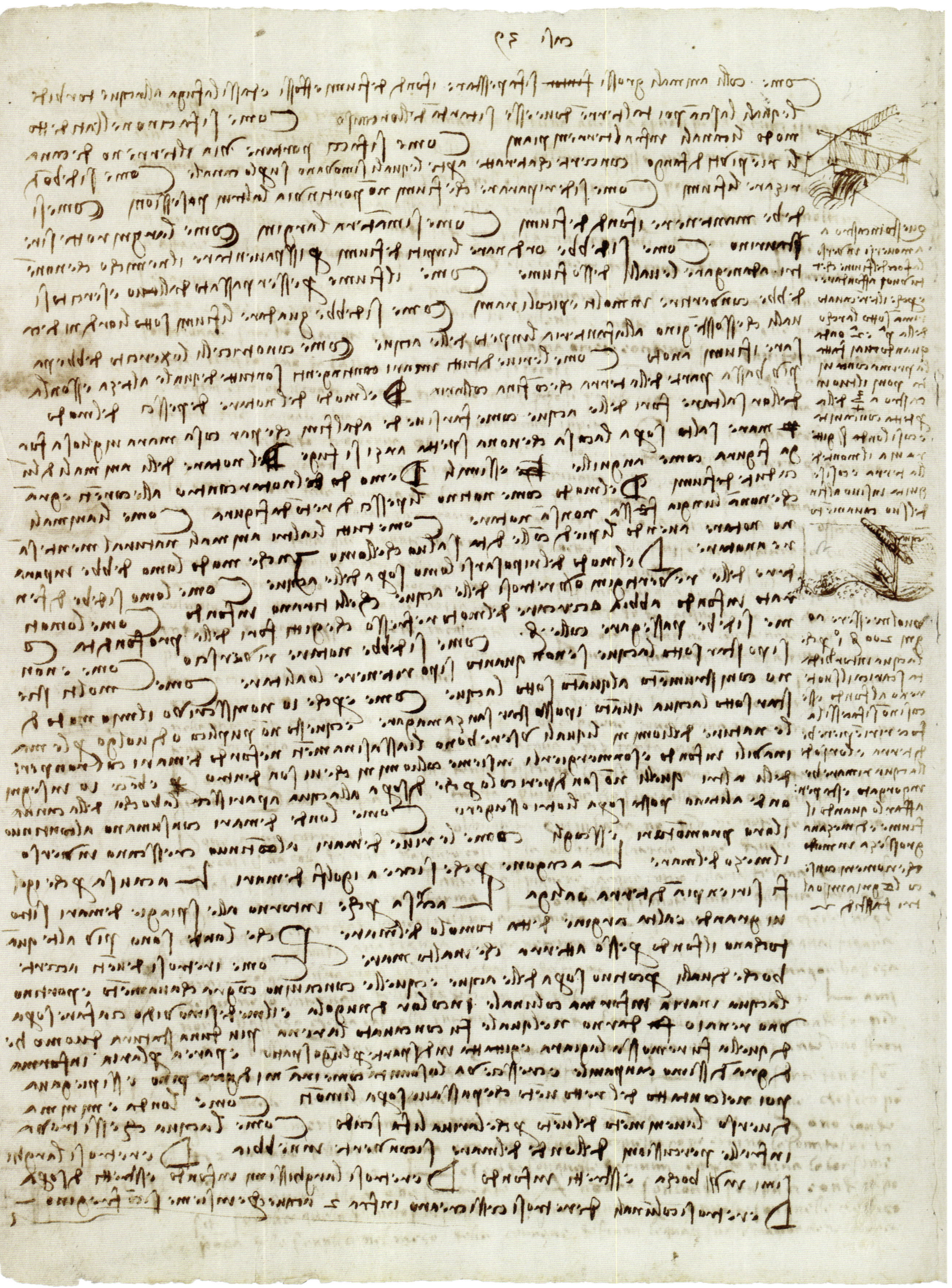

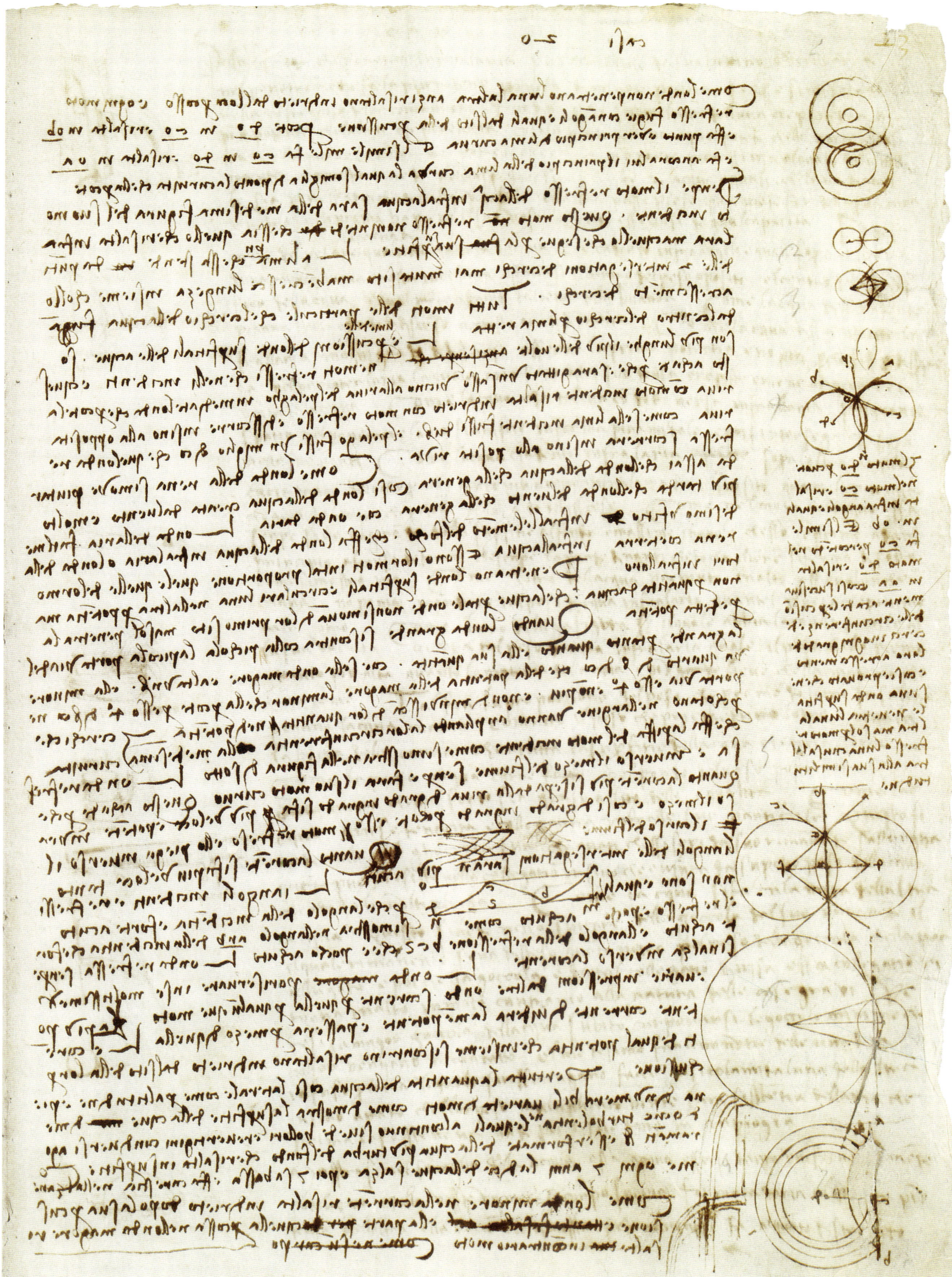

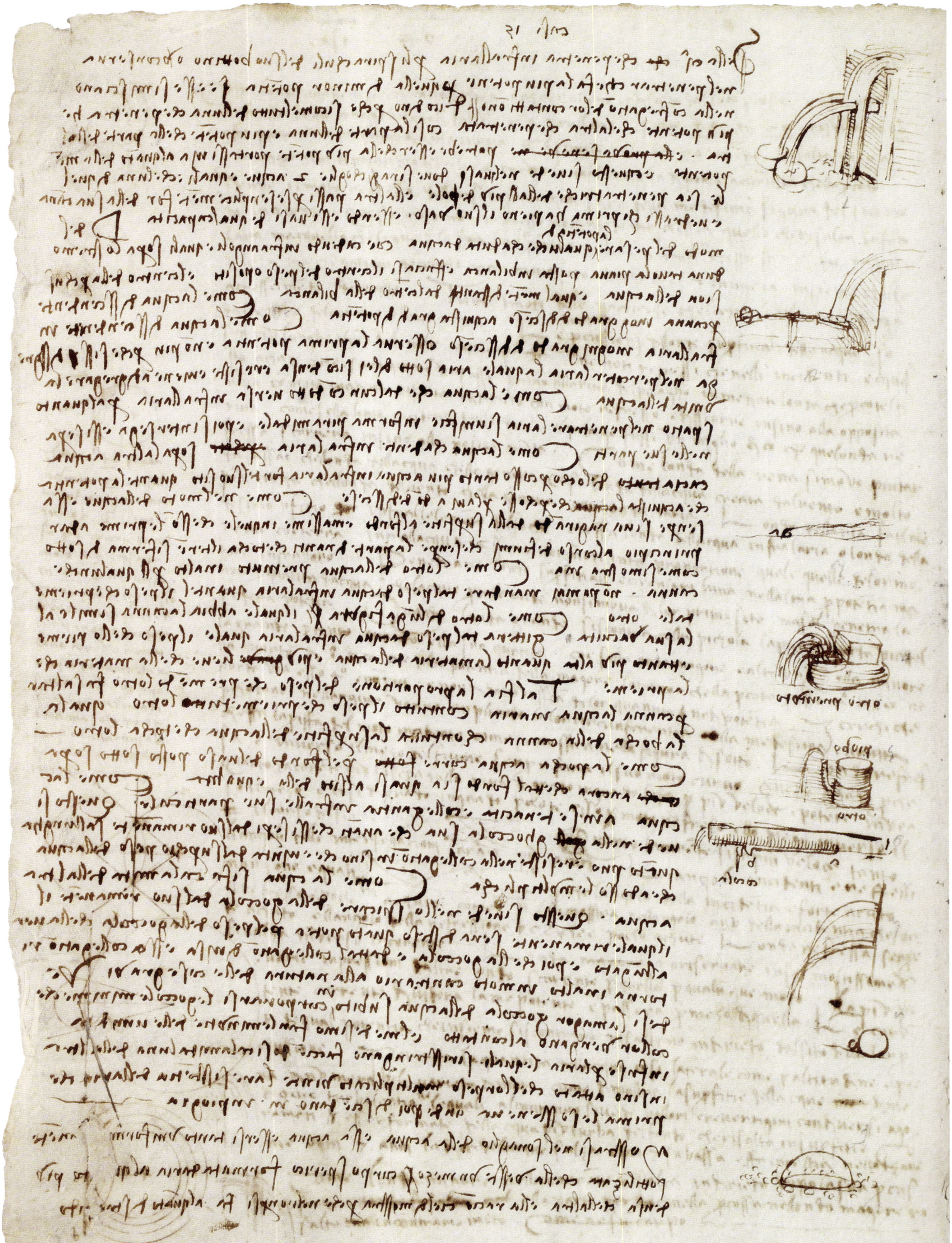

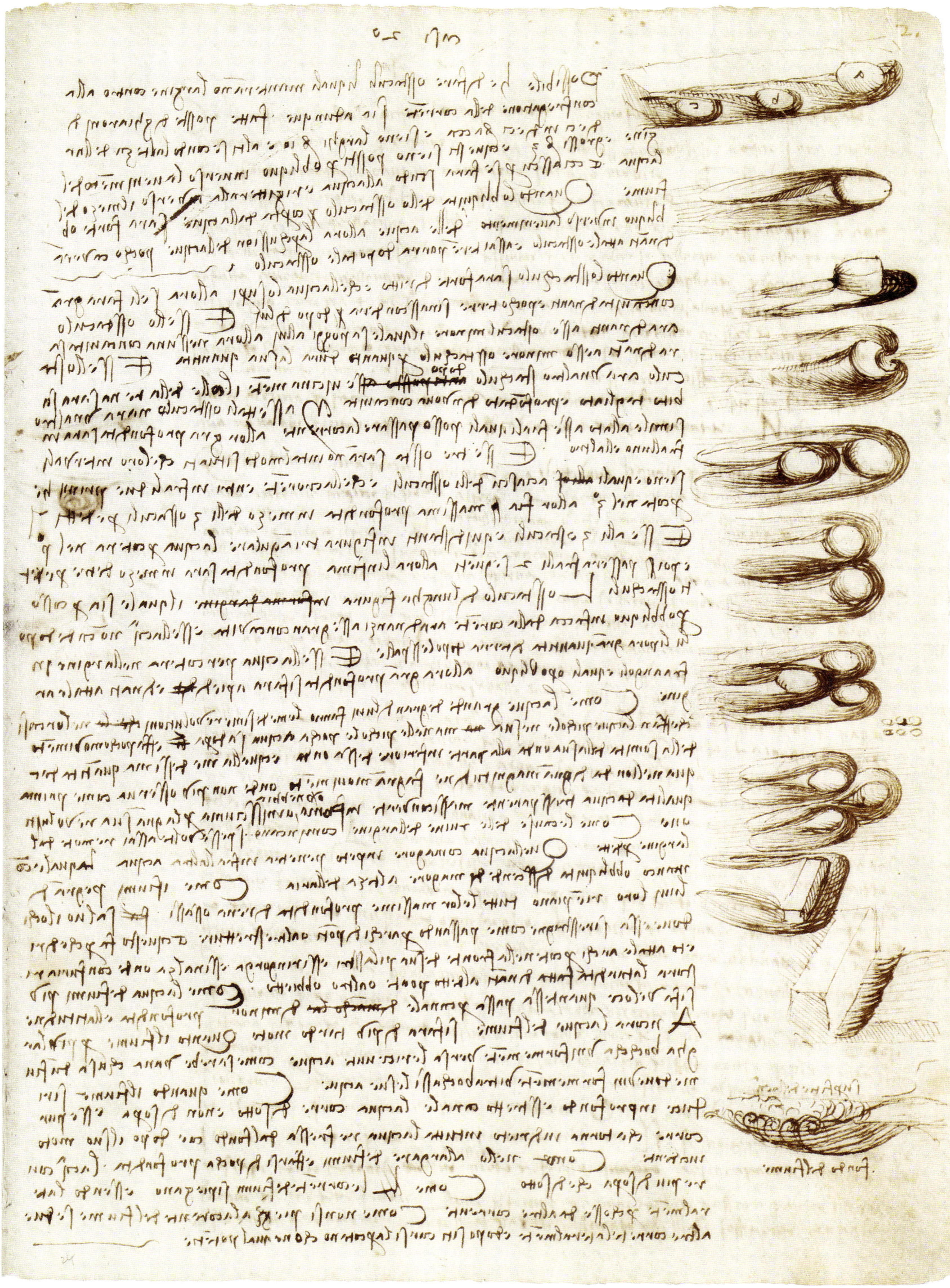

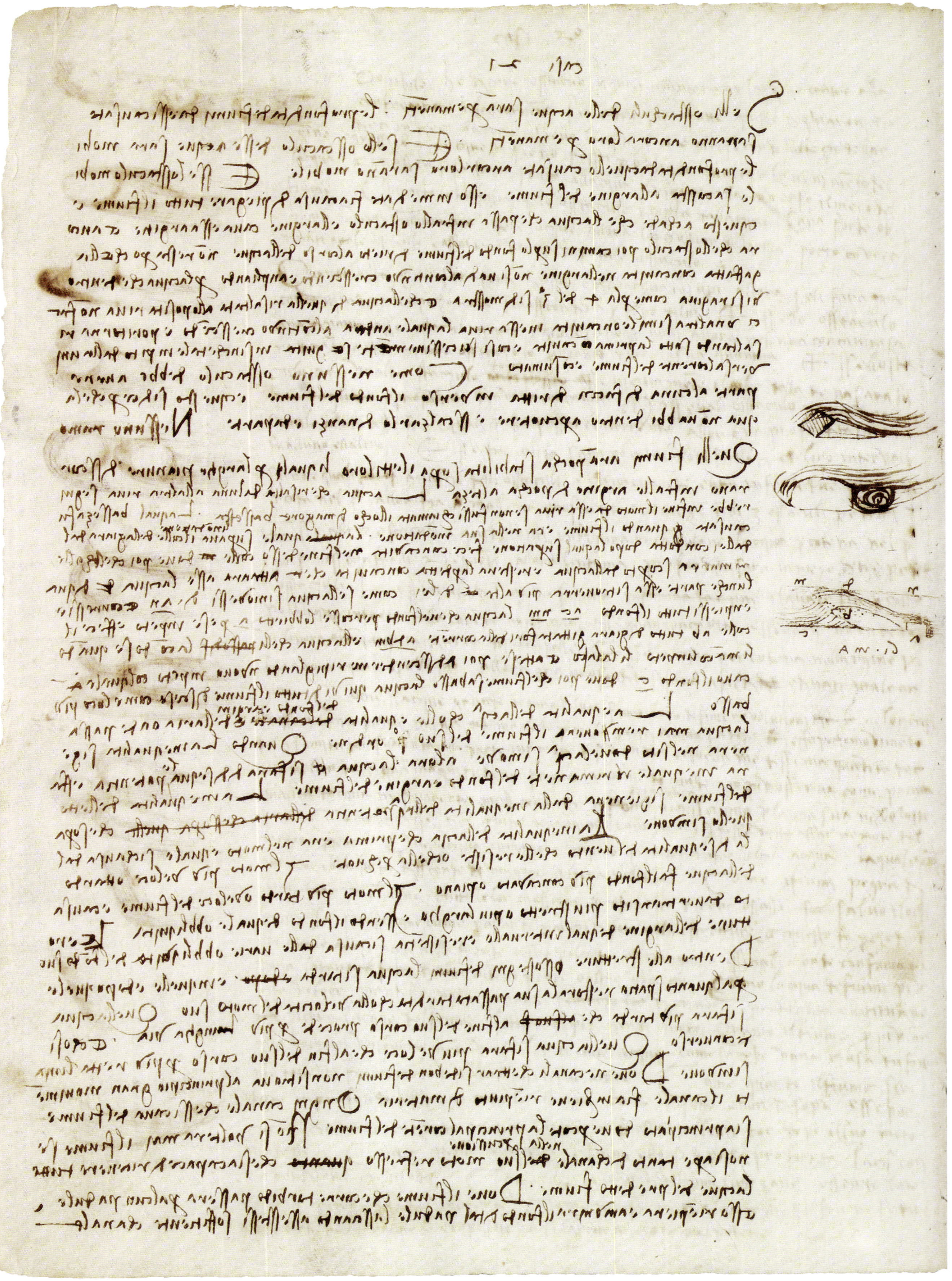

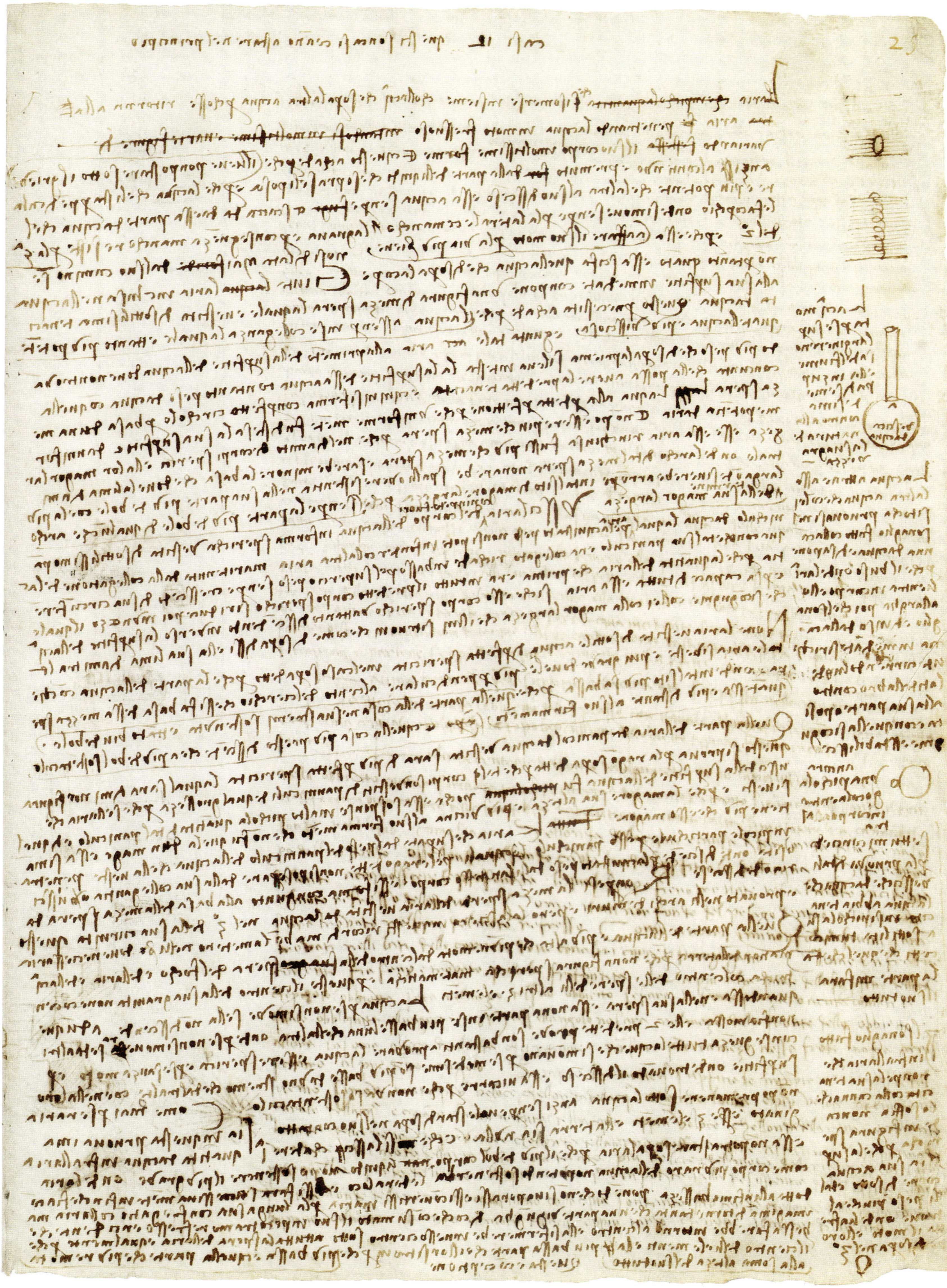

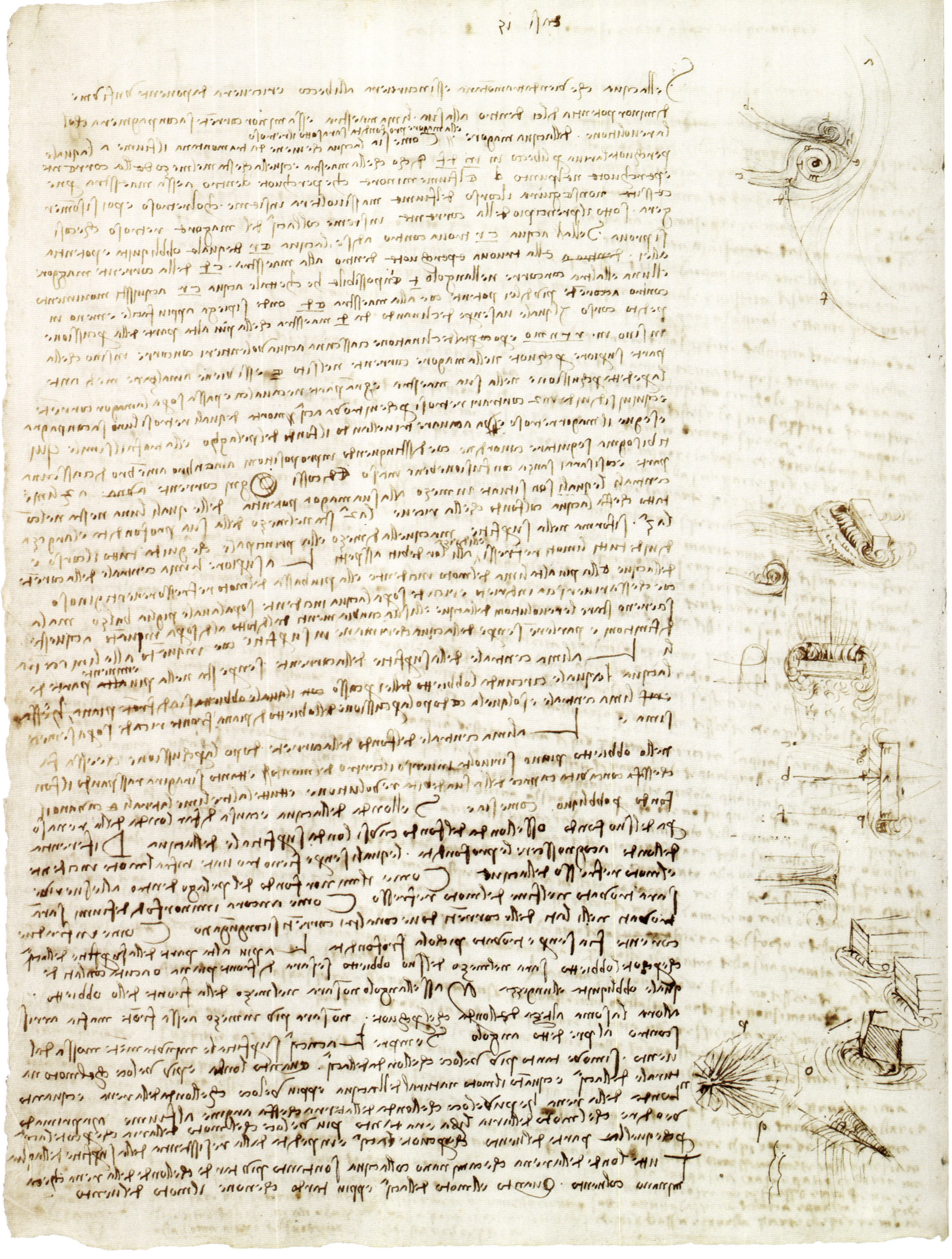

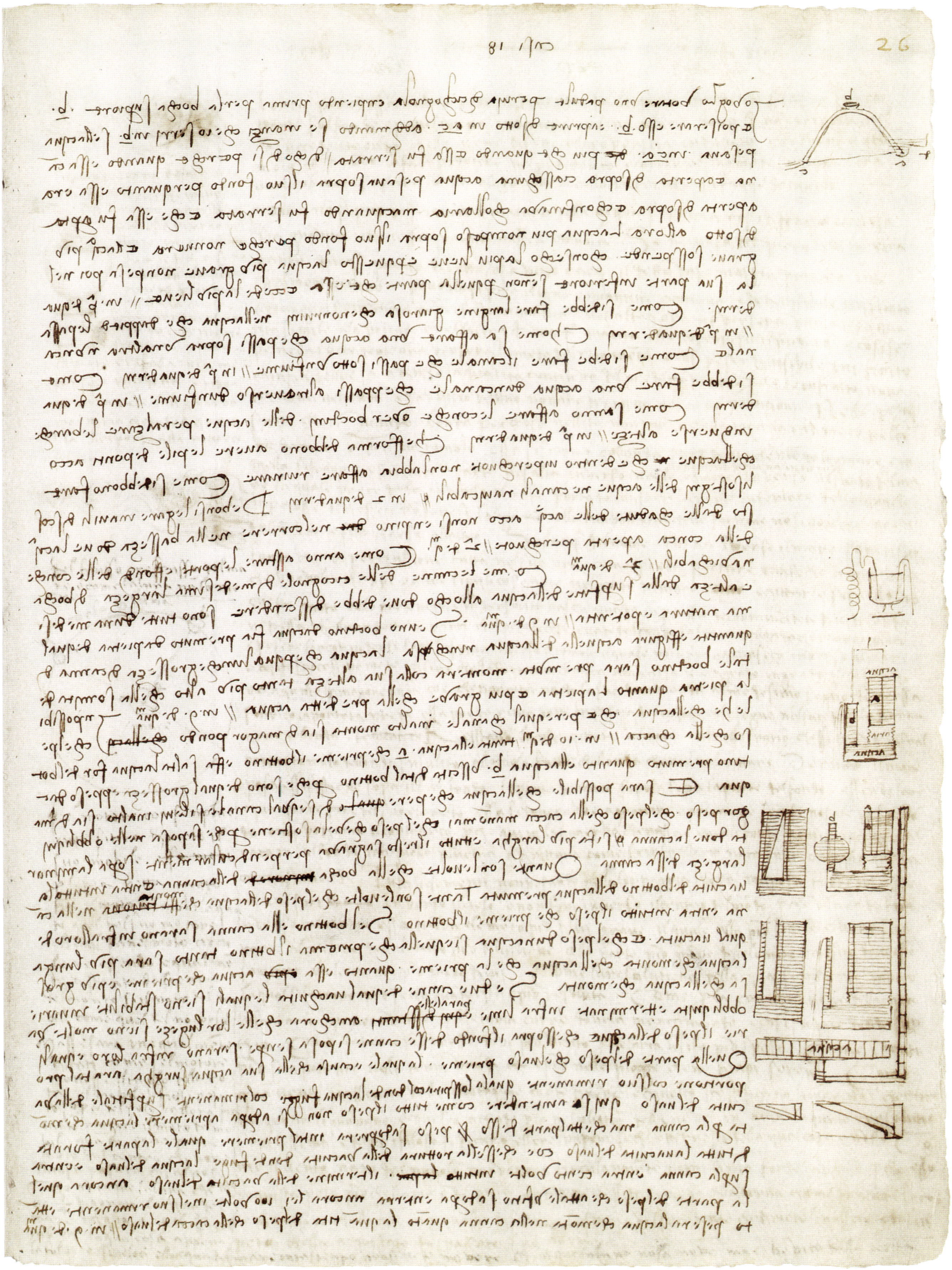

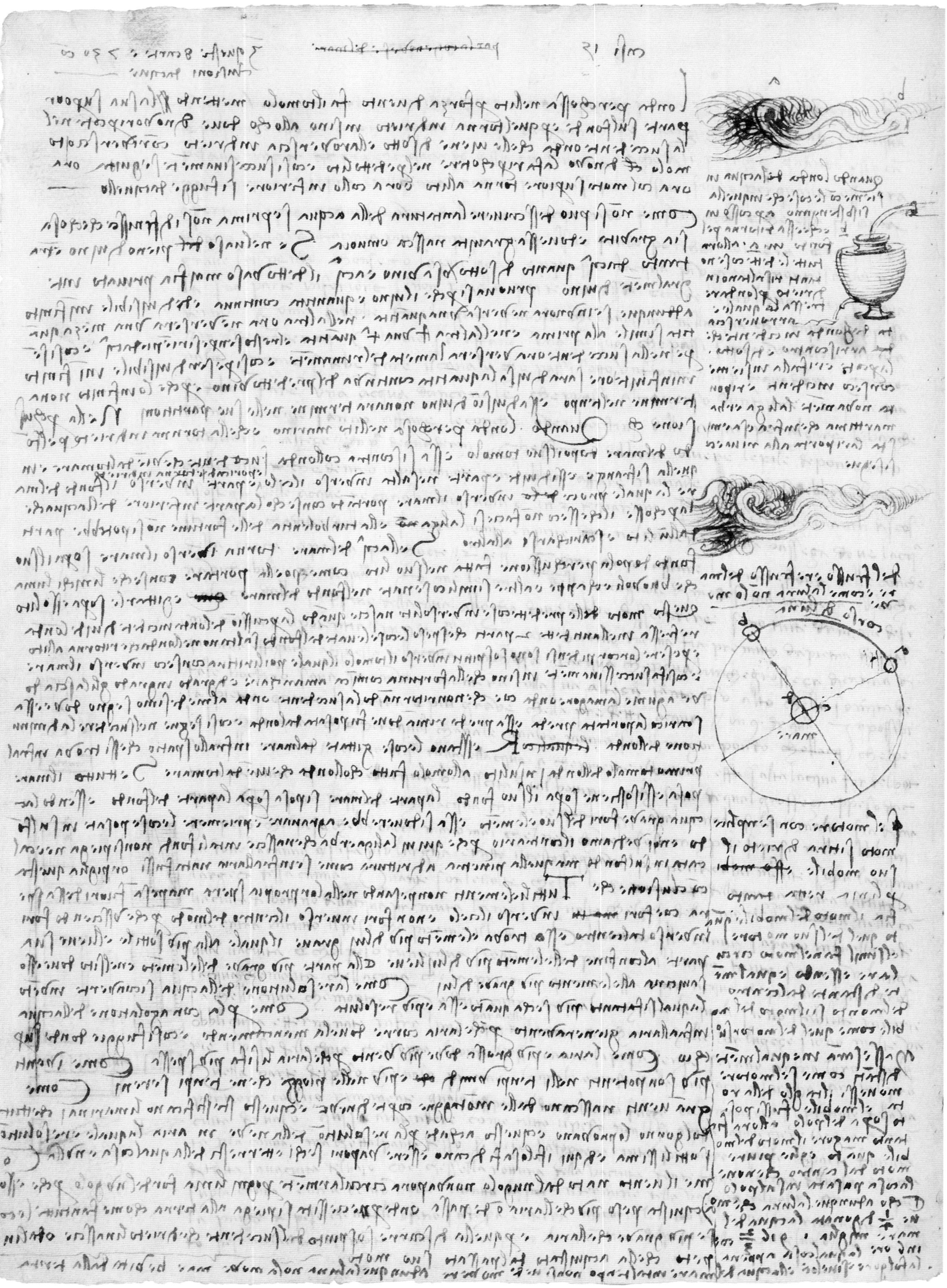

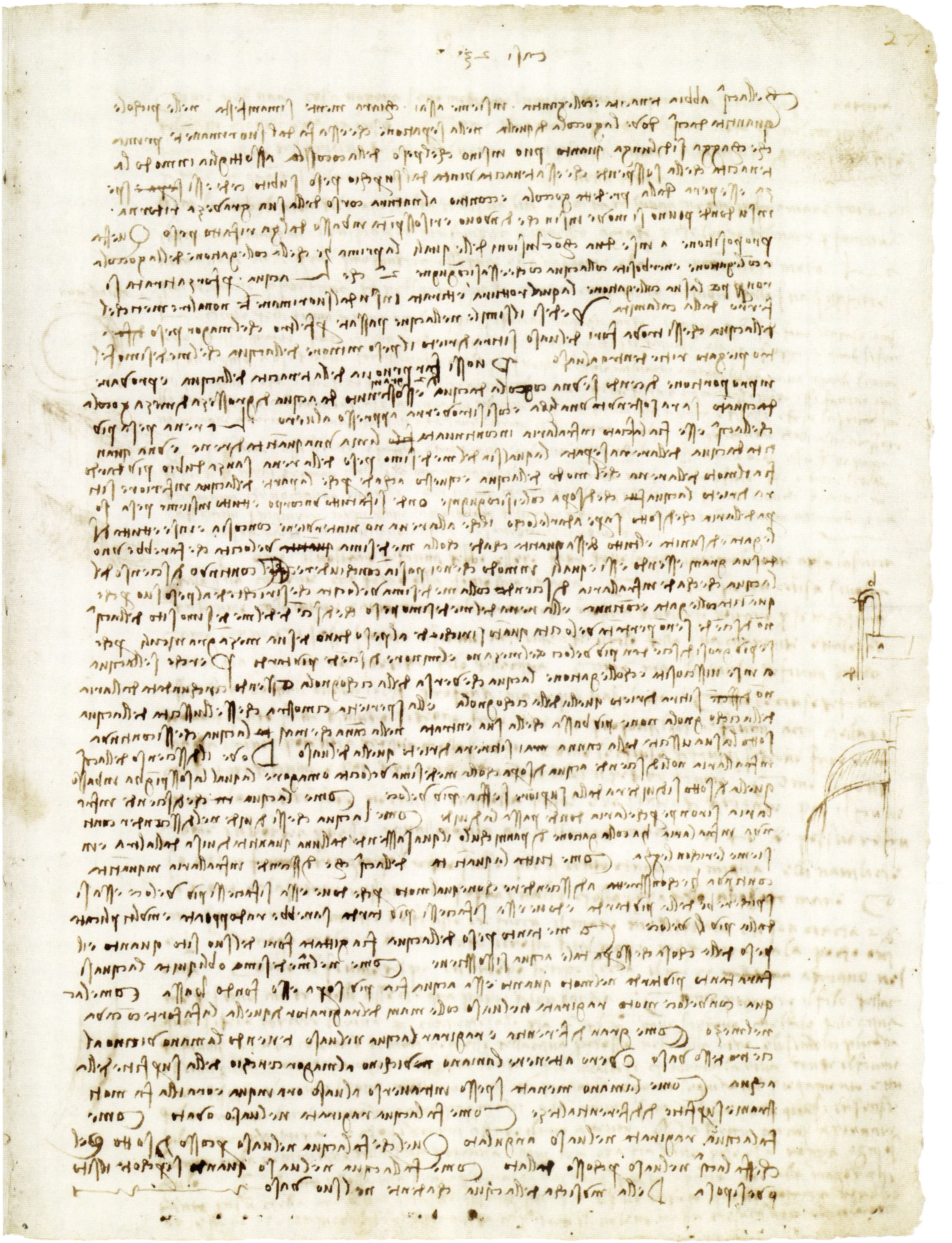

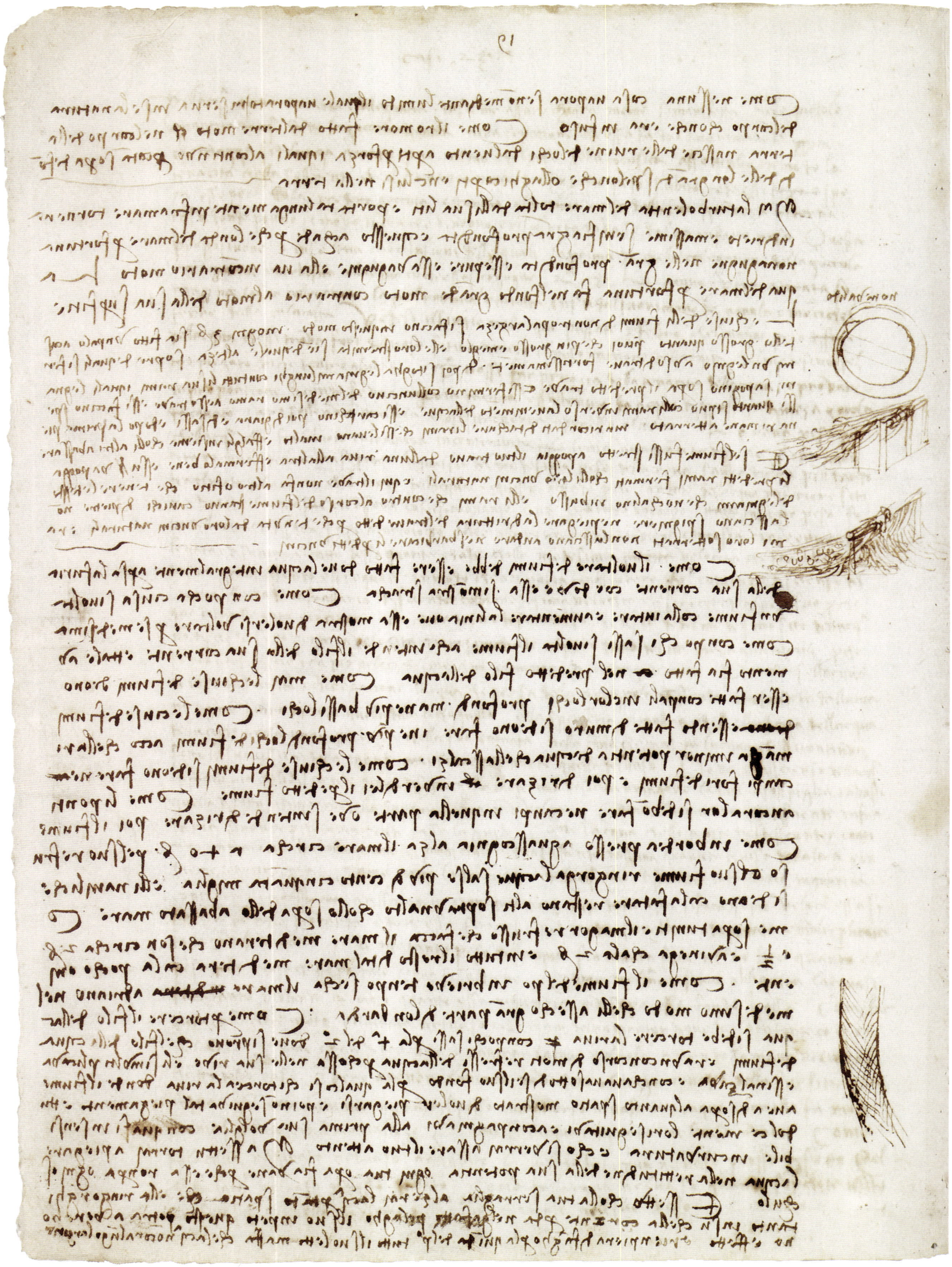

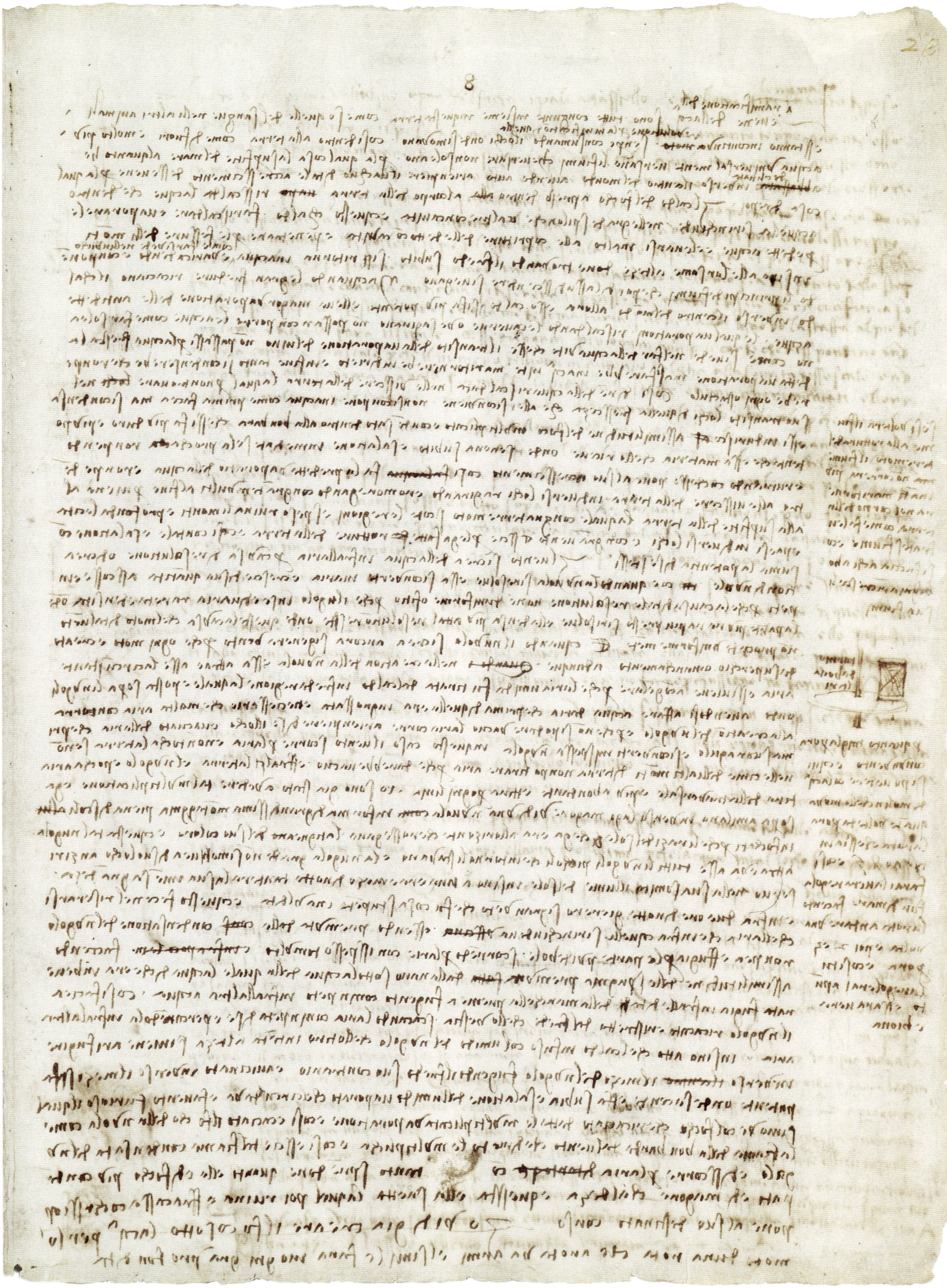

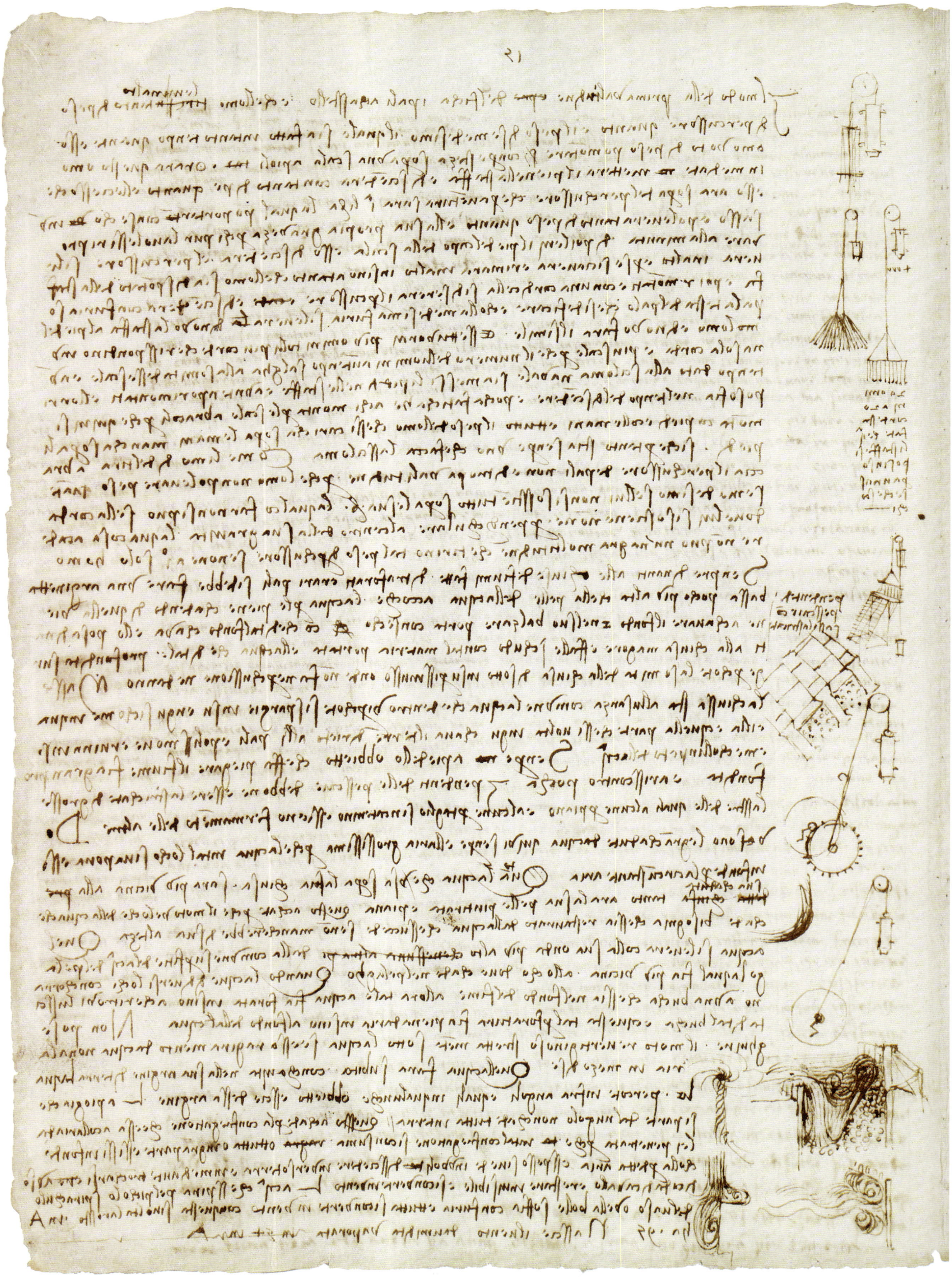

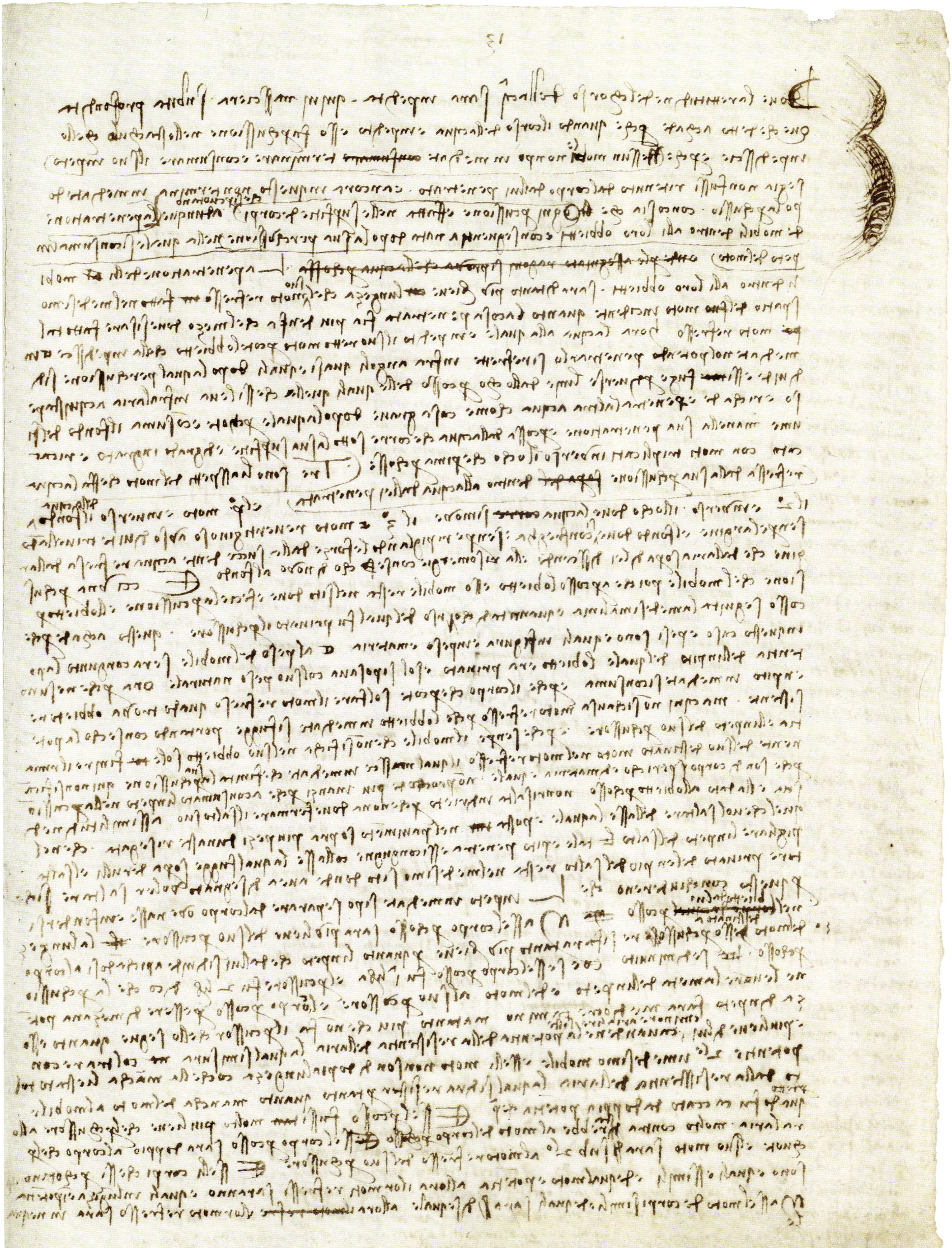

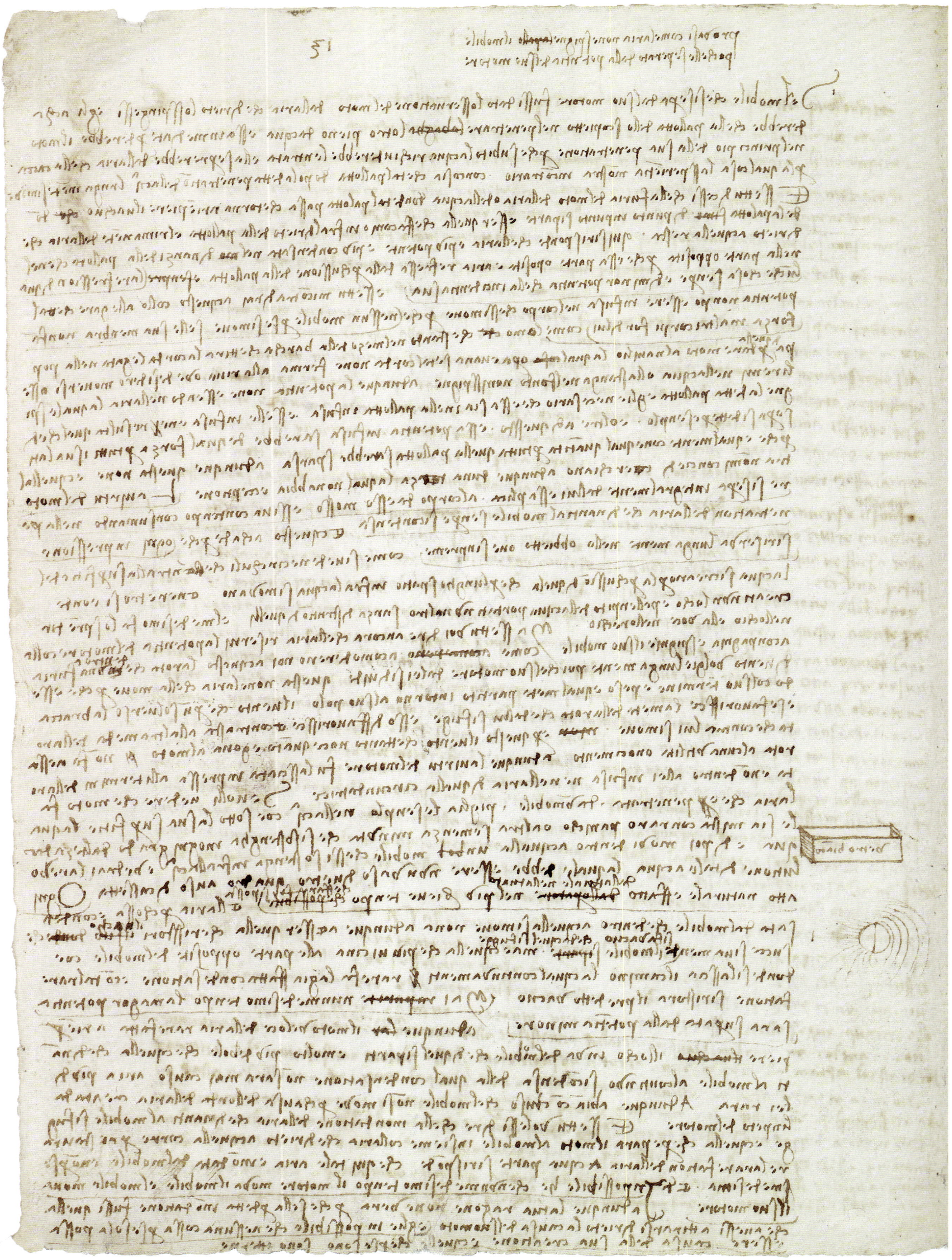

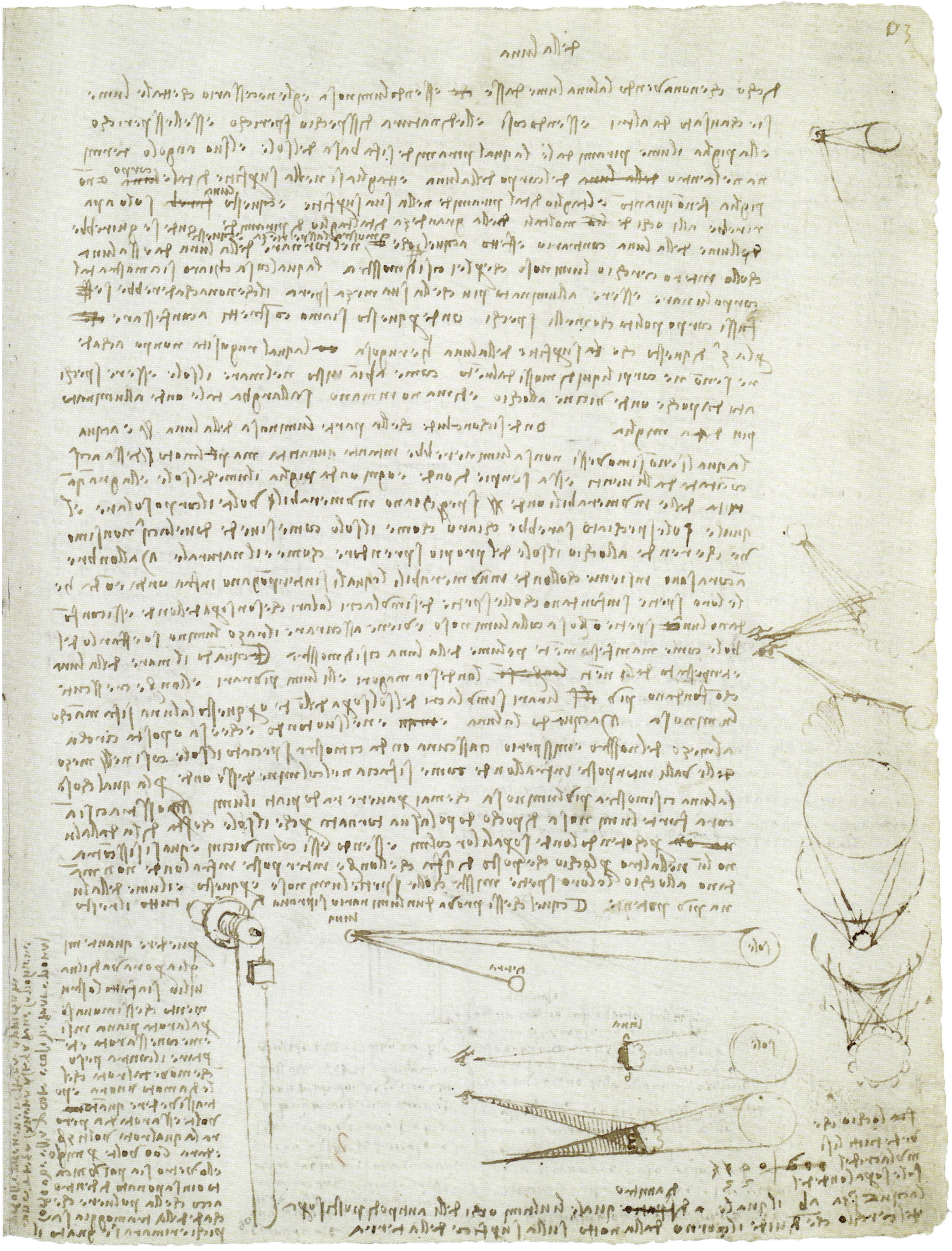

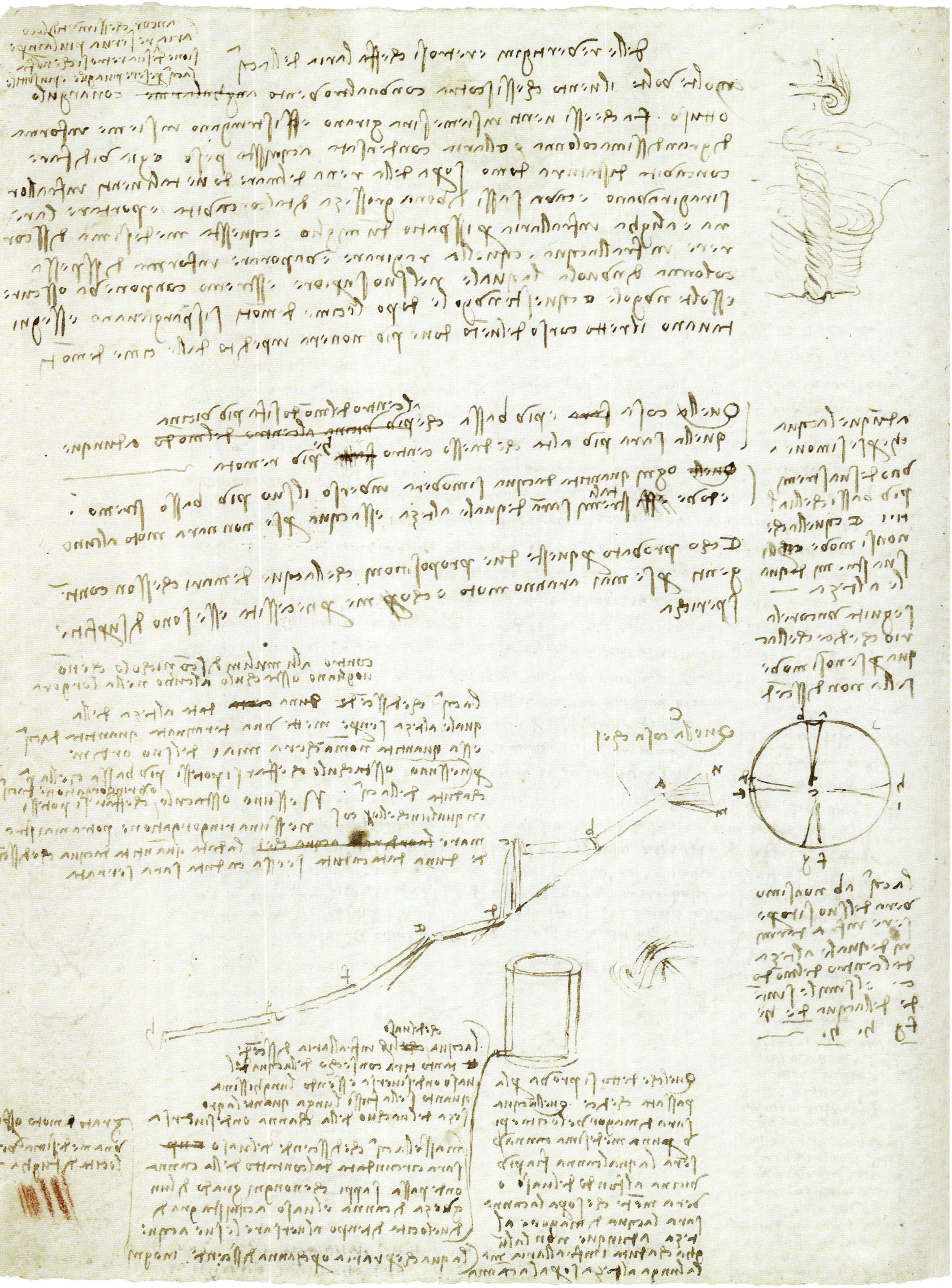

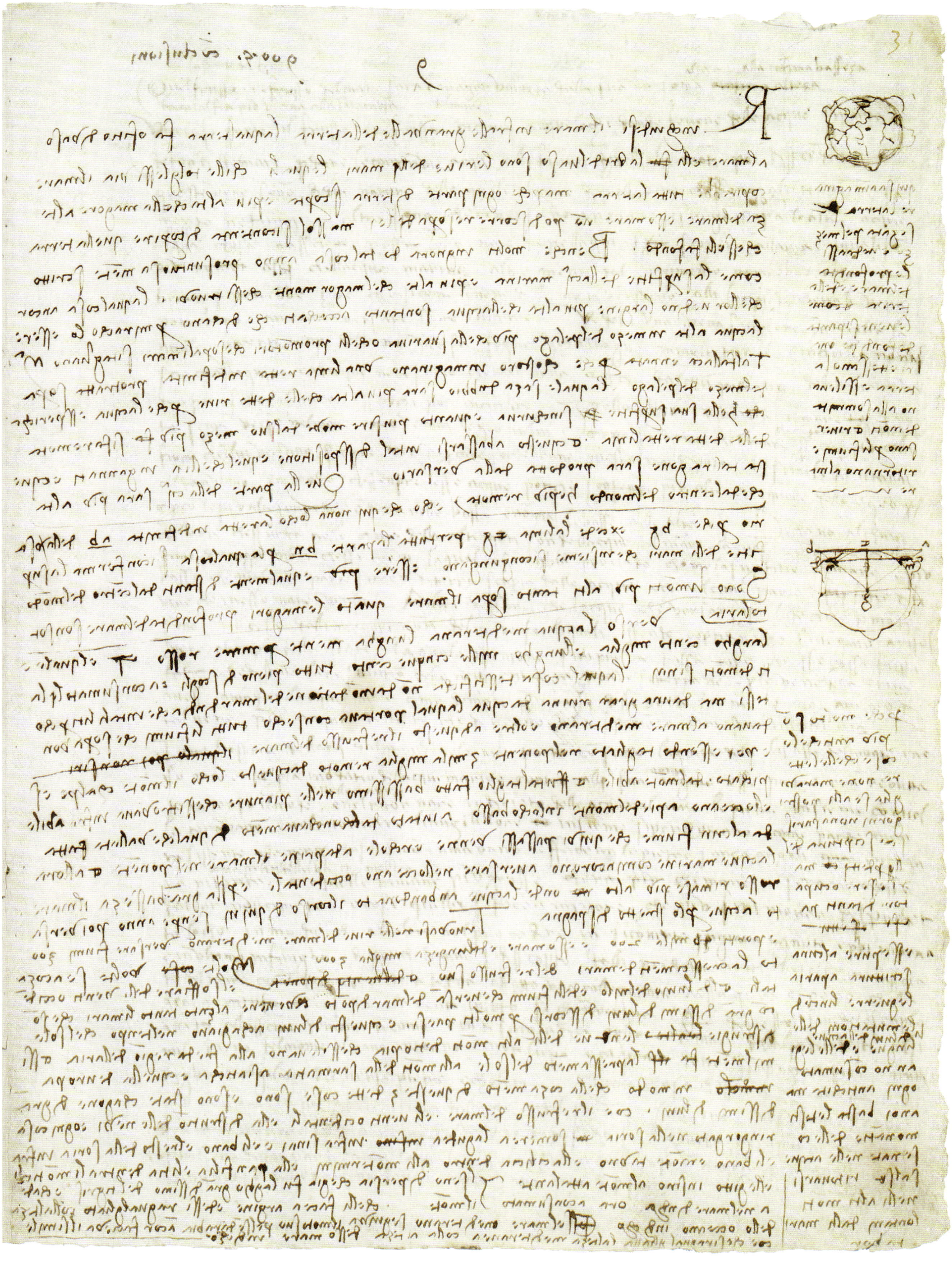

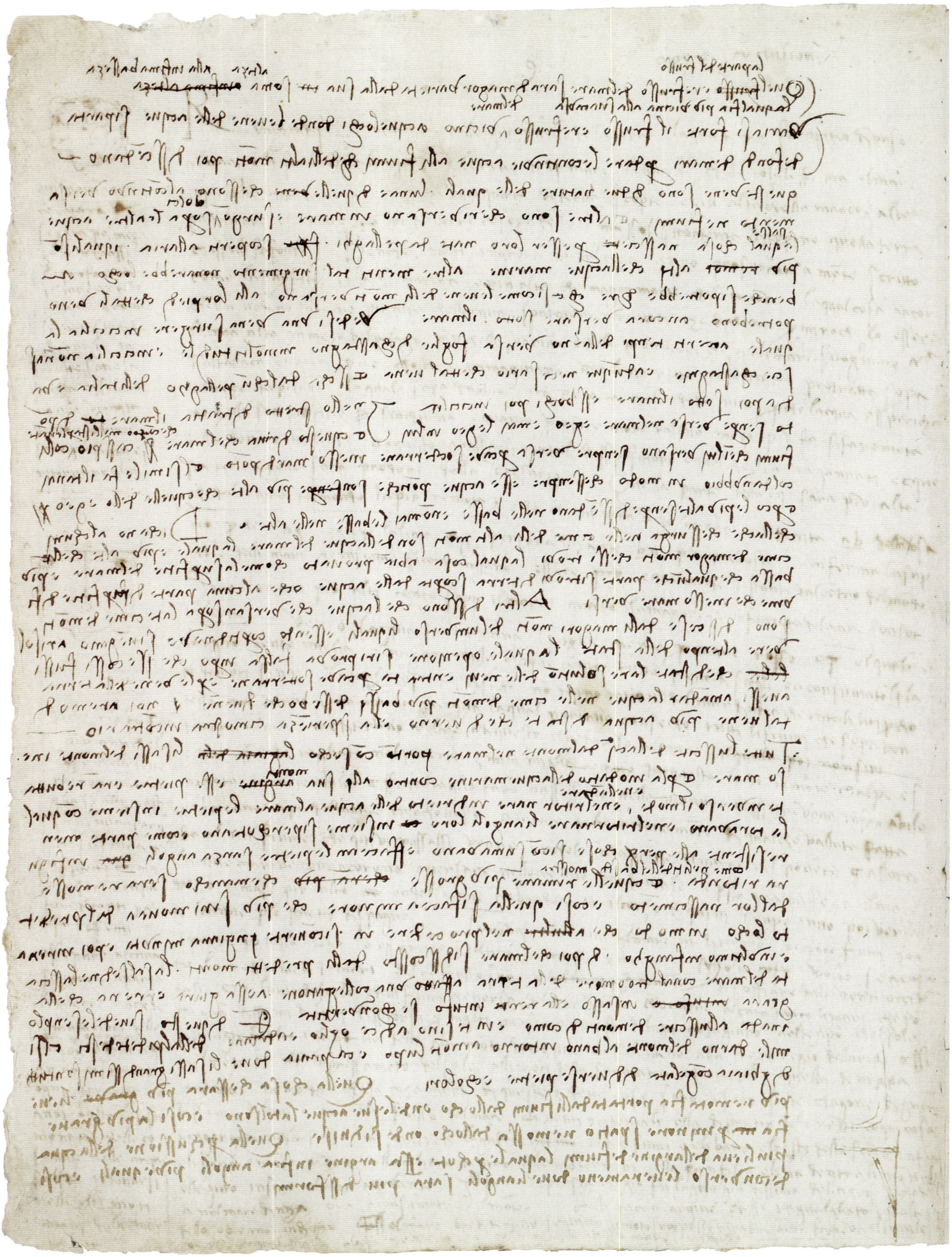

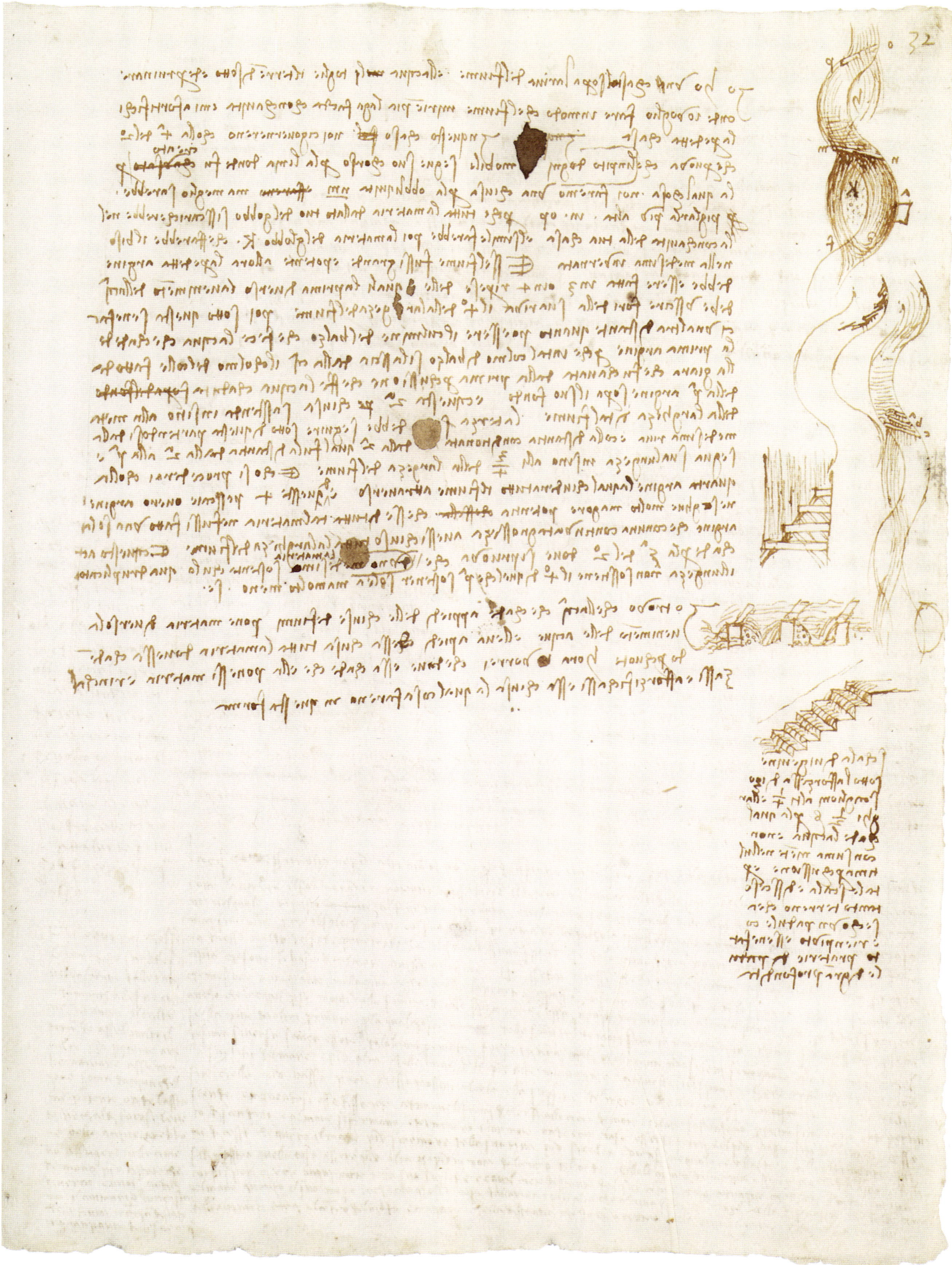

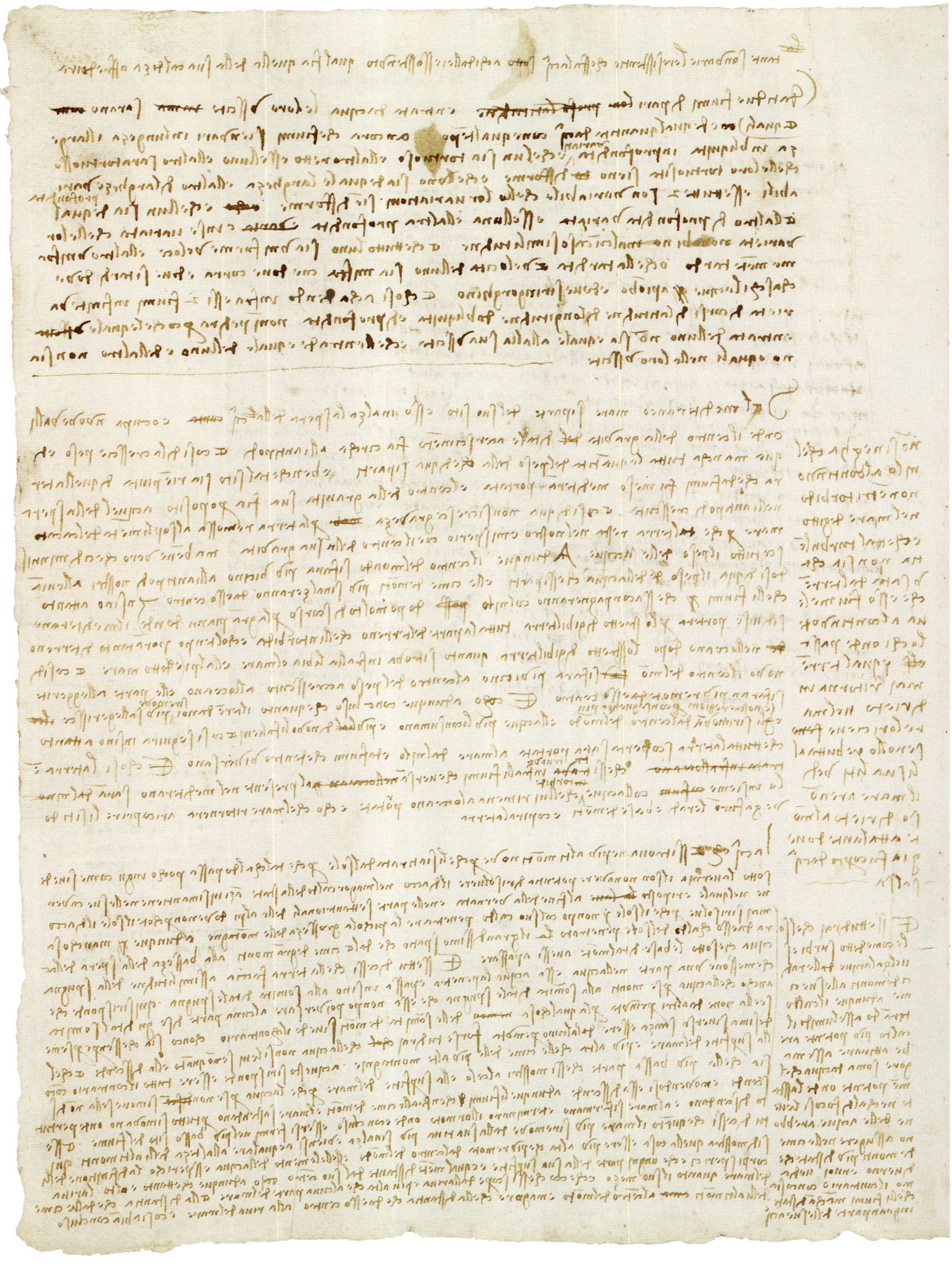

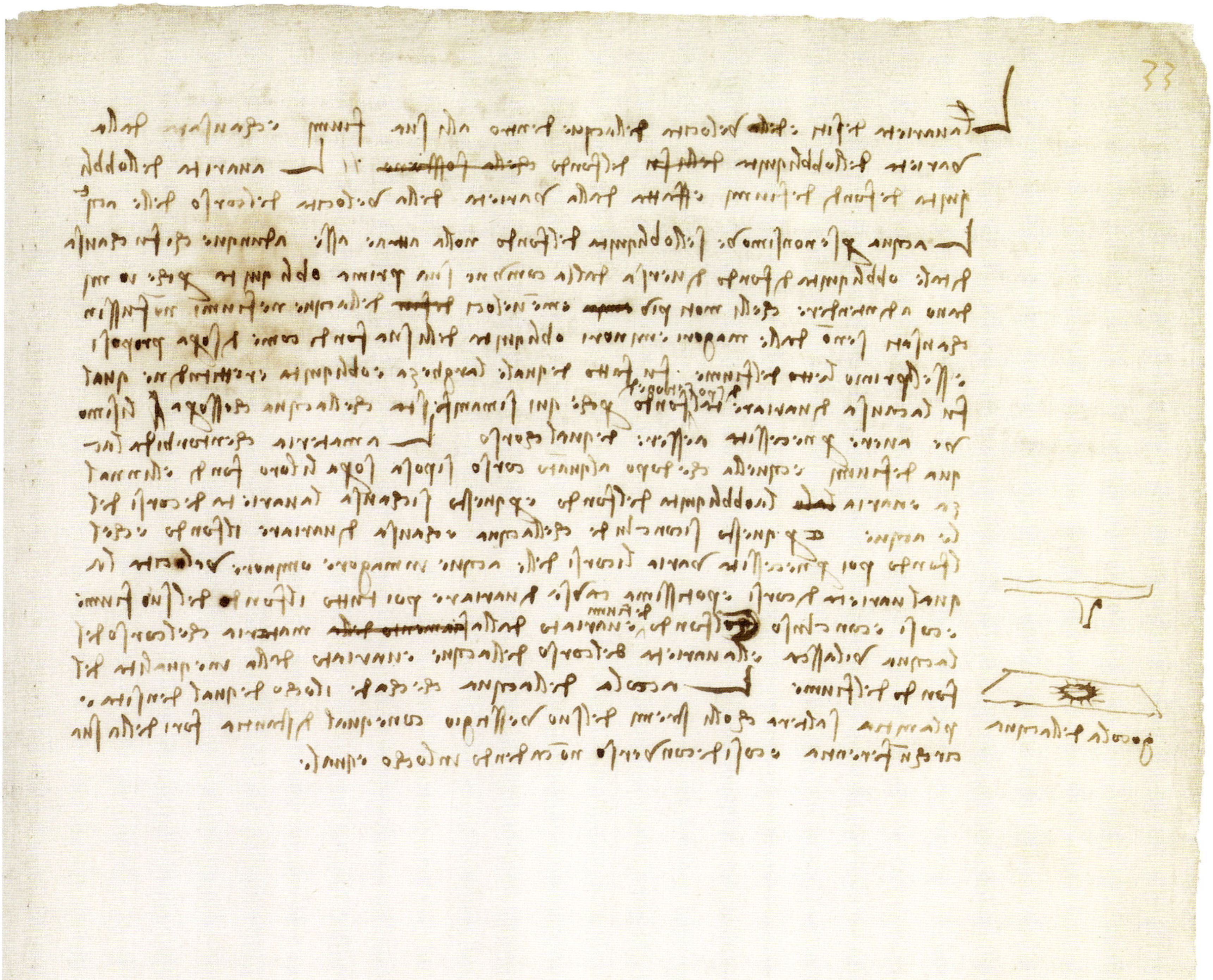

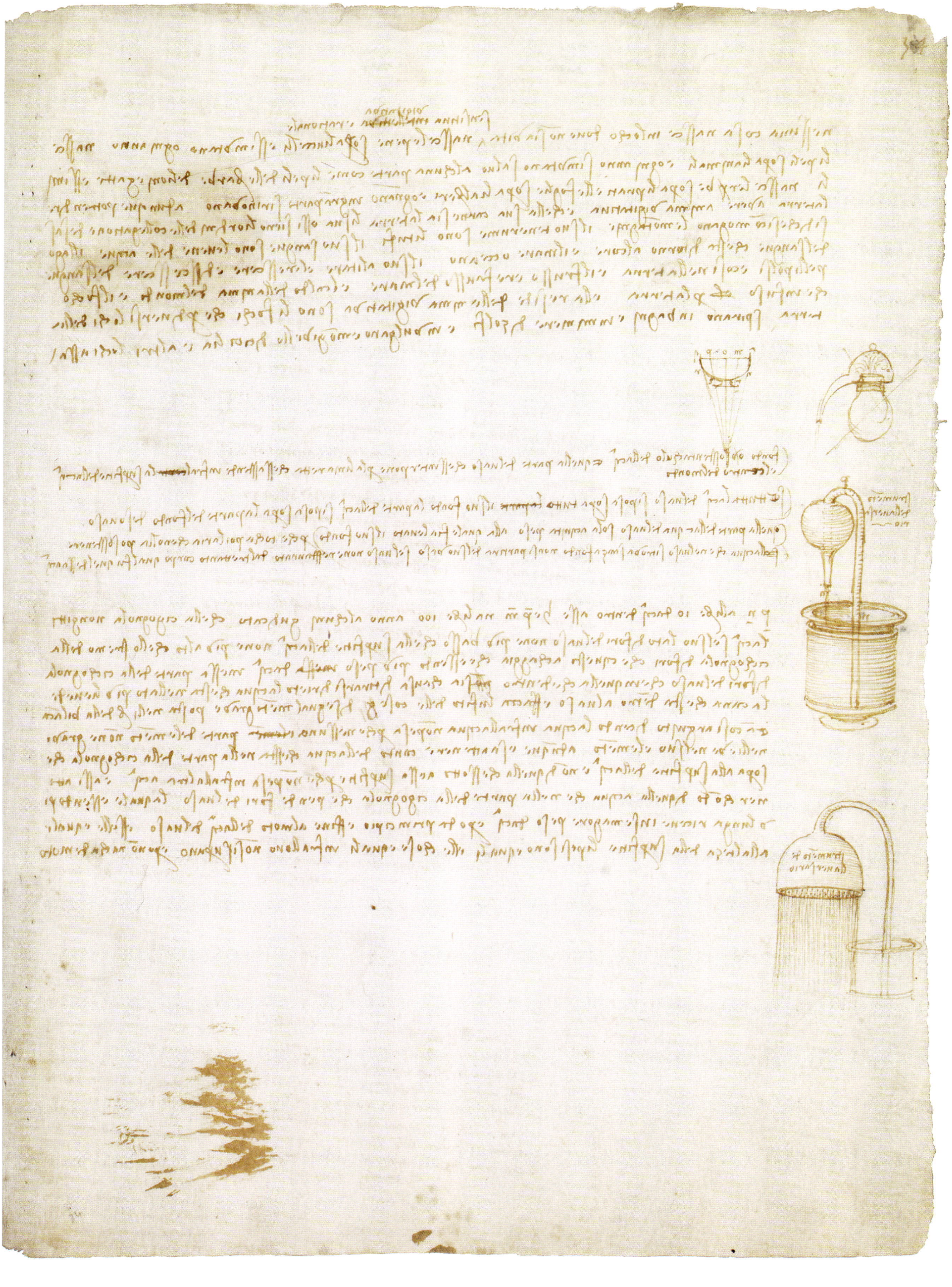

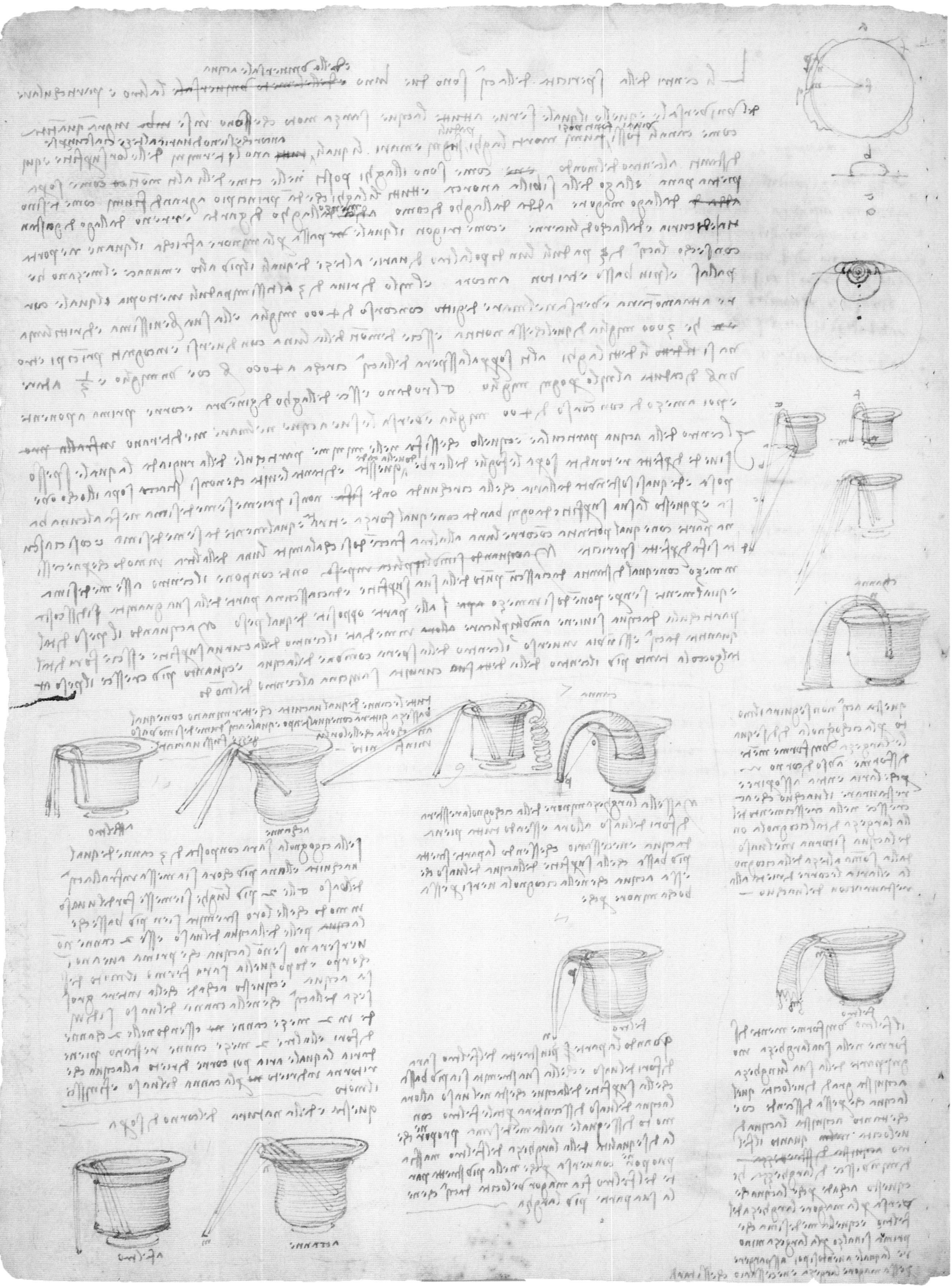

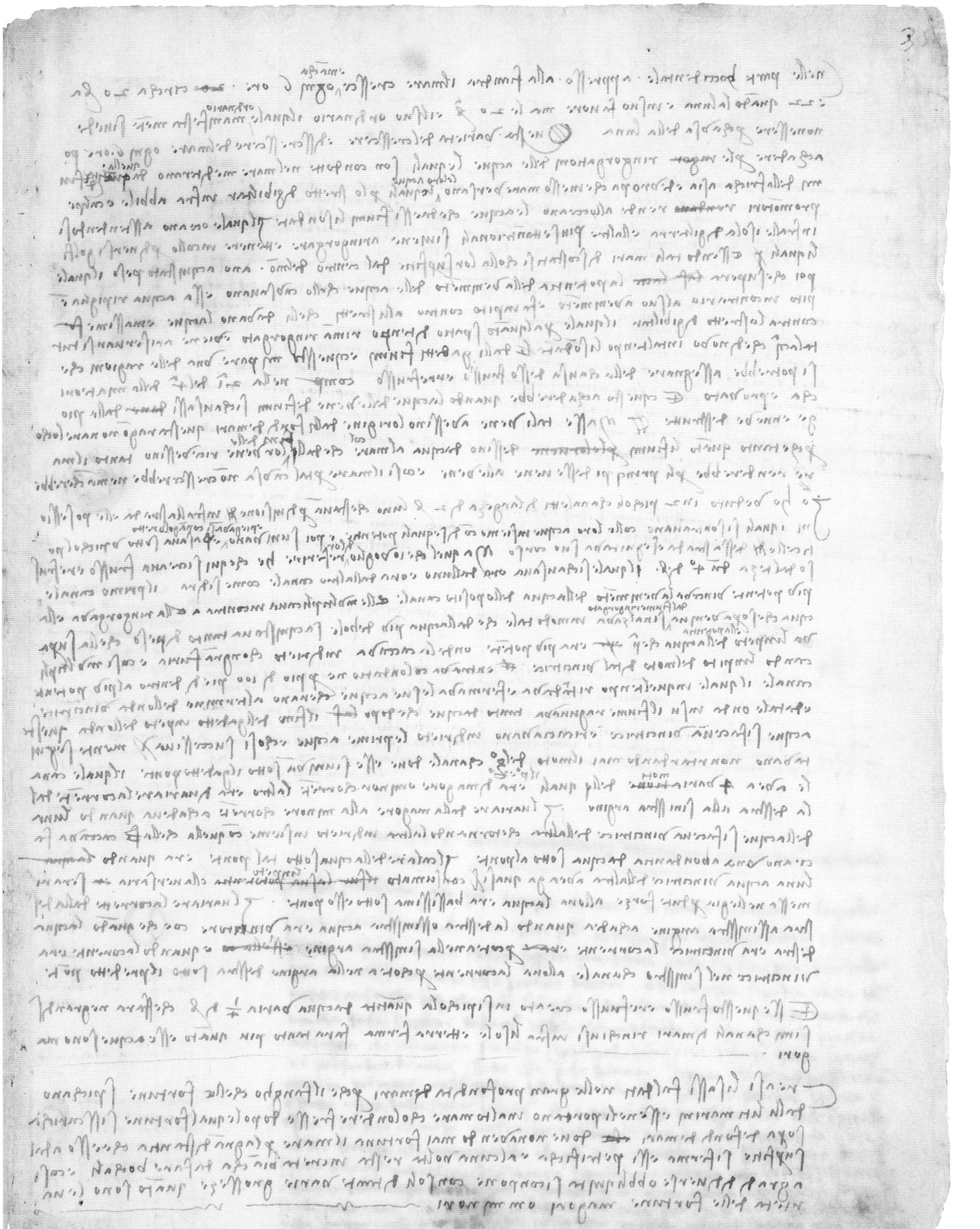

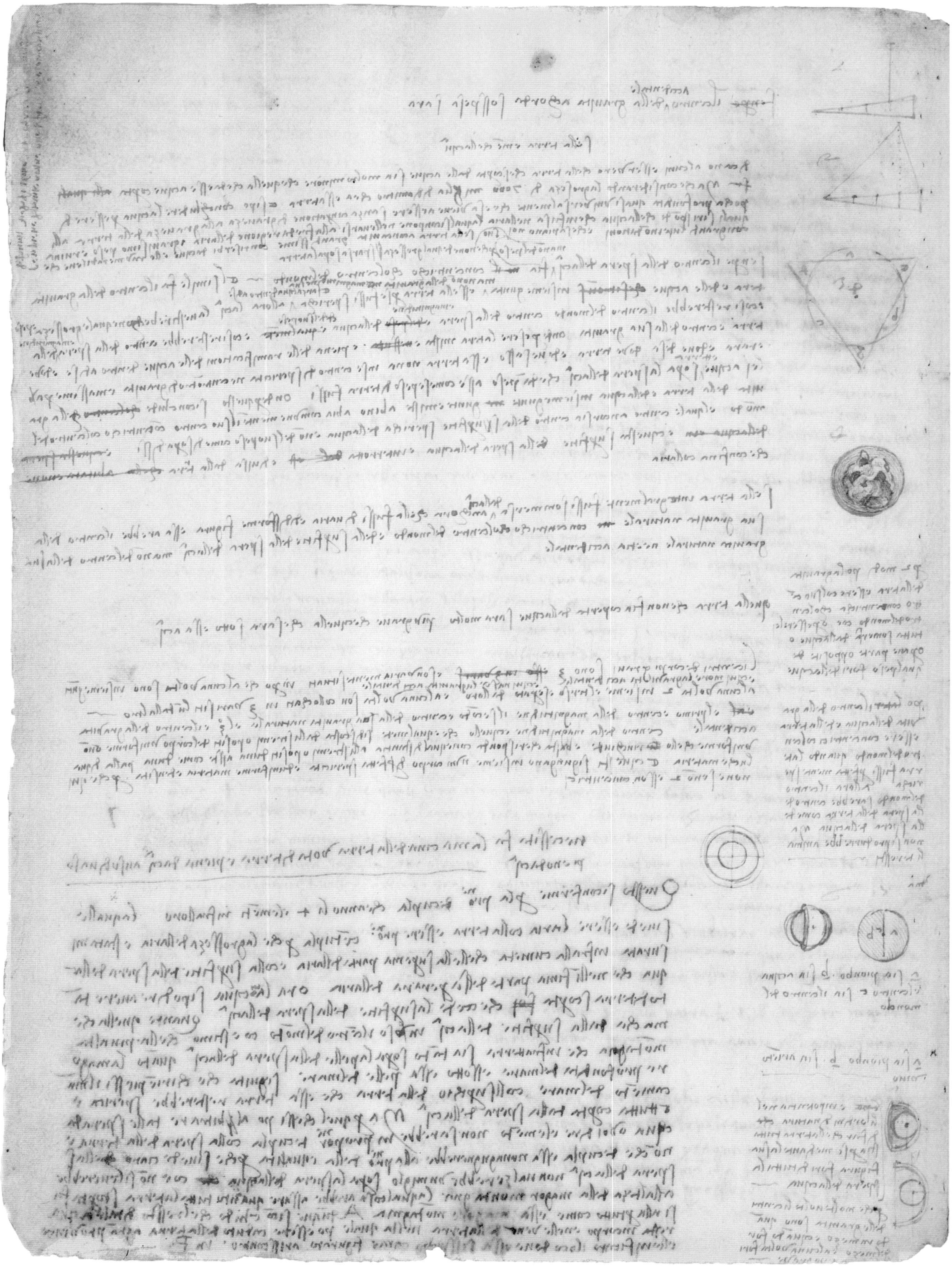

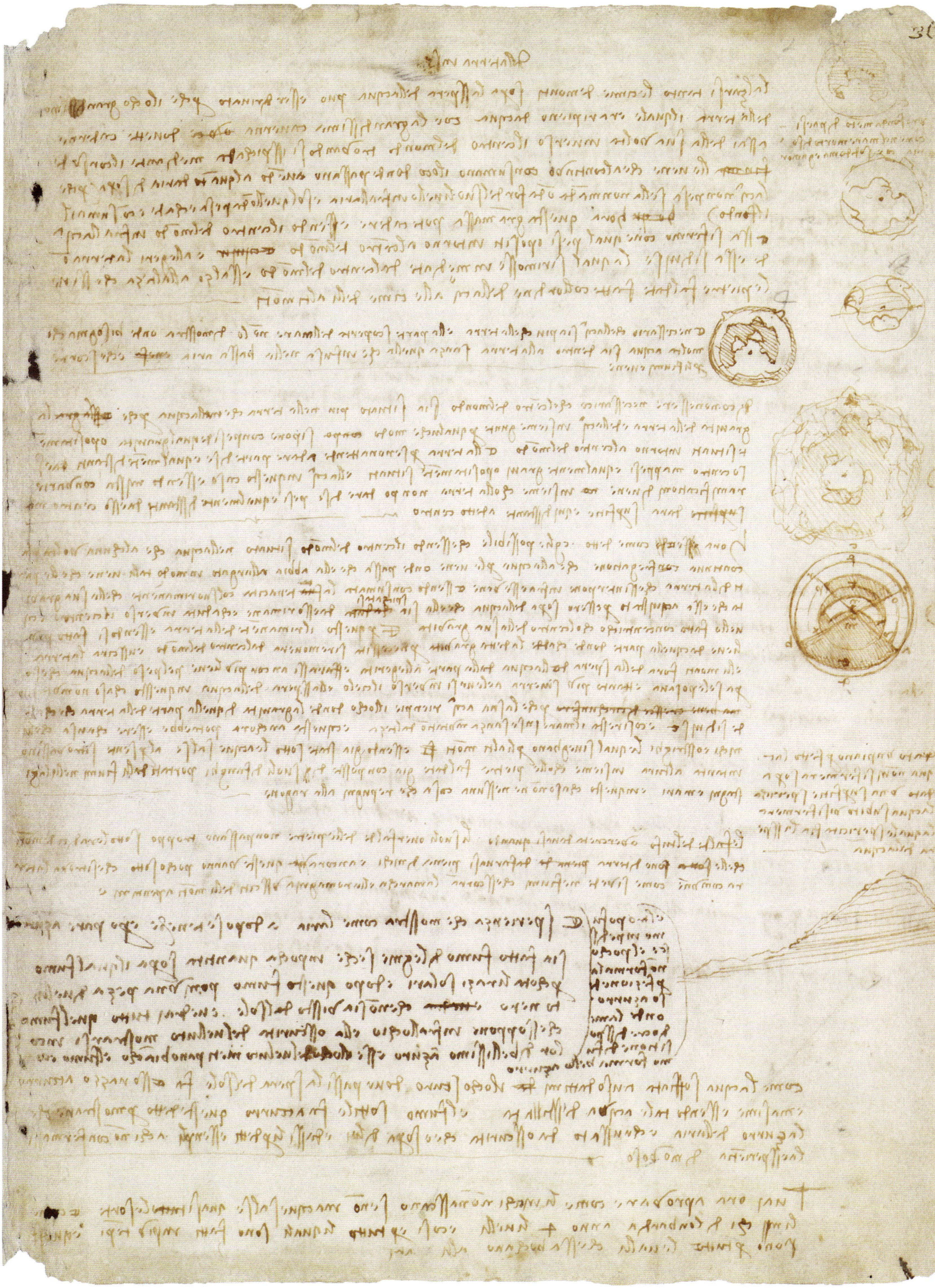

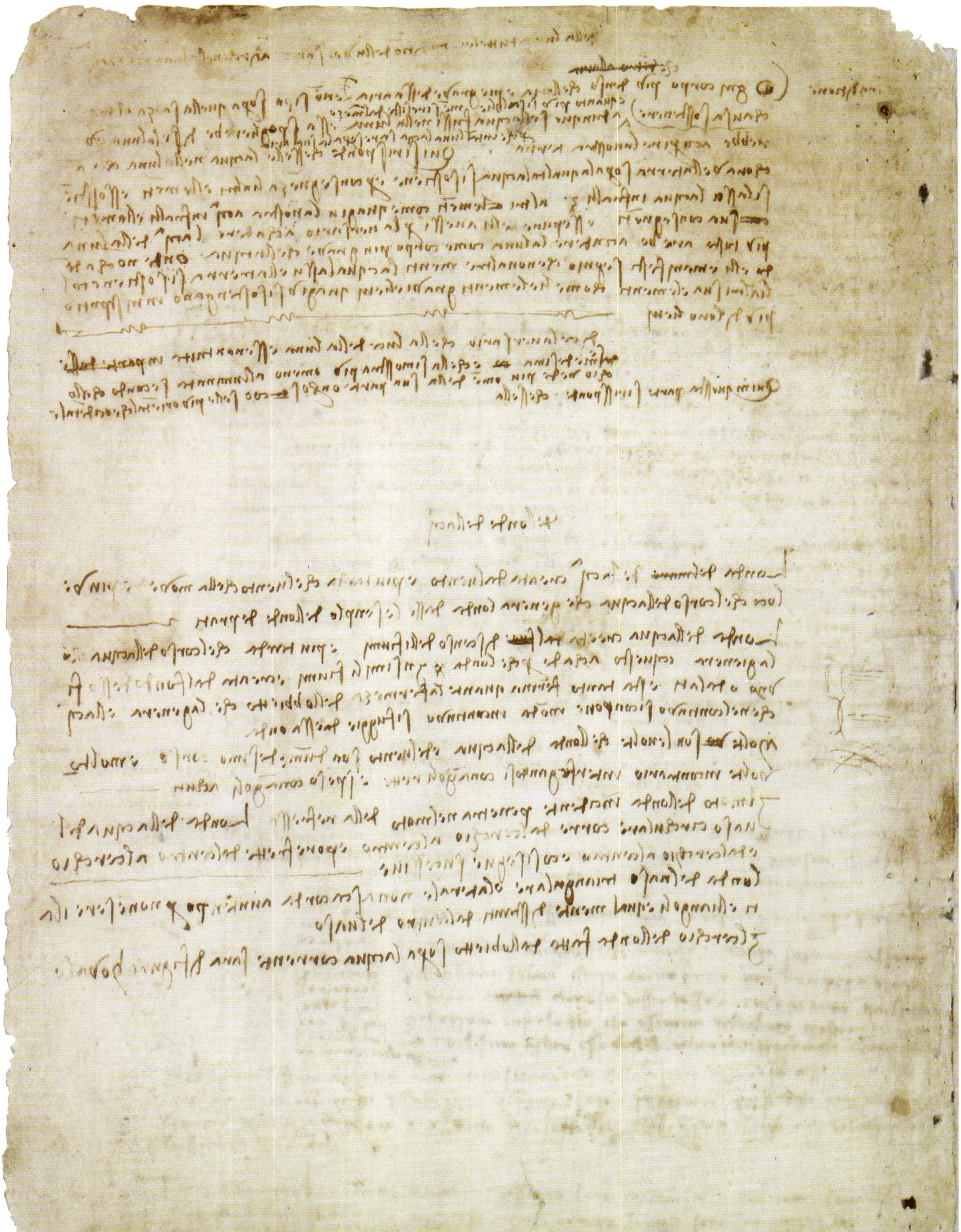

APPENDIX

PART 2

Ancient cover (sixteenth century?), recto (see Introduction)

Ancient cover (sixteenth century?), verso (see Introduction)

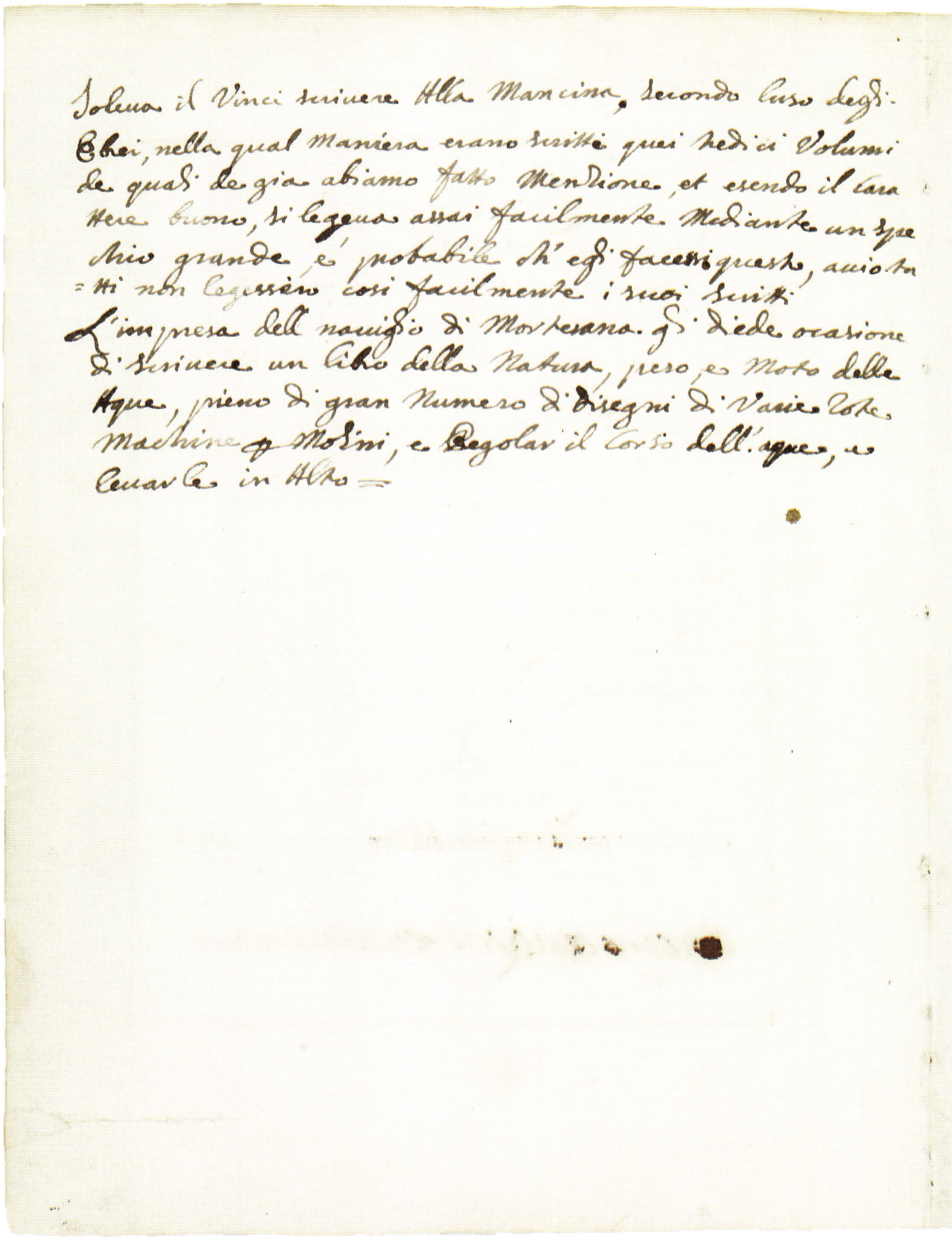
Soleua il Vinci scriuere Alla Mancina, secondo l'uso degli
Ebrei, nella qual Maniera erano scritti quei sedici Volumi
de quali se gia abiamo fatto Mentione, et esendo il cara
ttere buono, si legeua assai facilmente. Mediante un spe
chio grande è probabile ch' egli facessi questo, accio tu
=tti non legessero cosi facilmente i suoi scritti.
L'impresa dell nauiglio di Mortesana gli diede ocasione
di scriuere un libro della Natura, peso, e Moto delle
Aque, pieno di gran Numero di disegni di Varie Rote,
Machine per Molini, e Regolar il Corso dell'aque, e
leuarle in Alto =

Flyleaf with early eighteenth-century note (see Introduction and Volume II)

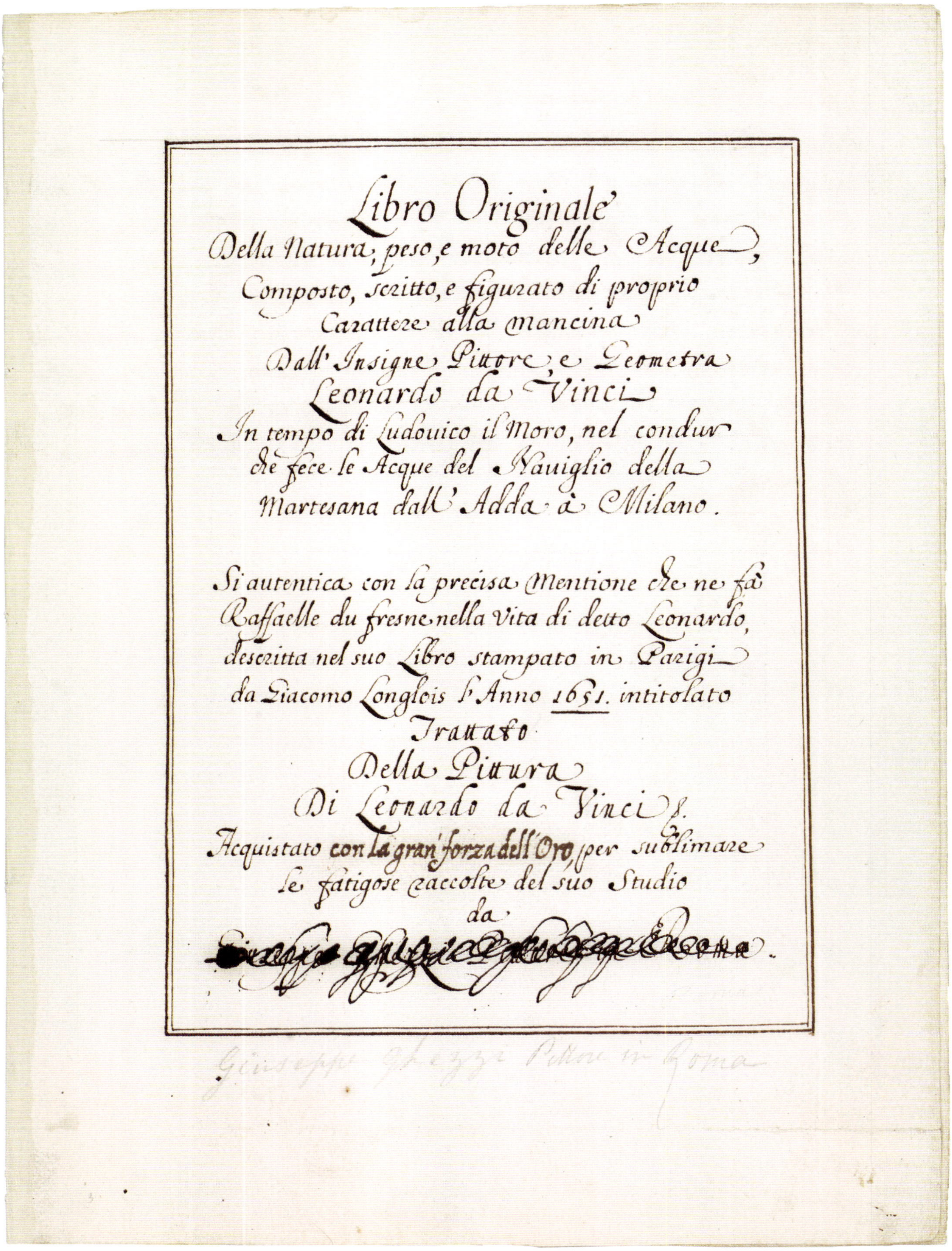

Libro Originale
Della Natura, peso, e moto delle Acque,
Composto, scritto, e figurato di proprio
Carattere alla mancina
Dall' Insigne Pittore, e Geometra
Leonardo da Vinci
In tempo di Ludouico il Moro, nel condur
che fece le Acque del Nauiglio della
Martesana dall' Adda à Milano.

Si autentica con la precisa Mentione che ne fà
Raffaelle du fresne nella Vita di detto Leonardo,
descritta nel suo Libro stampato in Parigi
da Giacomo Langlois l'Anno 1651. intitolato
Trattato
Della Pittura
Di Leonardo da Vinci &.
Acquistato con la gran forza dell'Oro per sublimare
le fatigose raccolte del suo Studio
da
[redacted].

Giuseppe Ghezzi Pittore in Roma

Flyleaf with early eighteenth-century title page (see Introduction and Volume II)

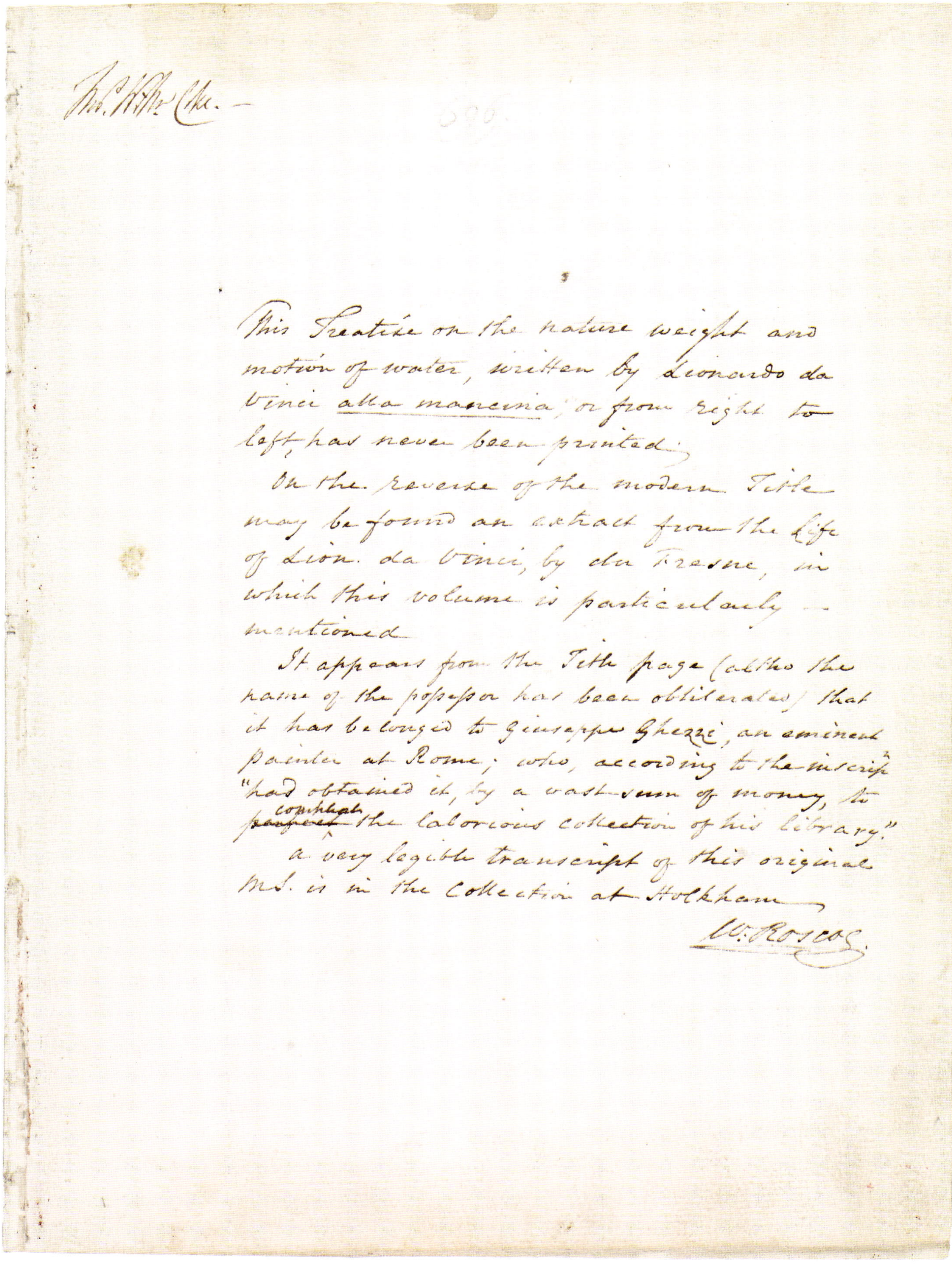

Thos. Wm. Coke.—

This Treatise on the nature weight and motion of water, written by Leonardo da Vinci alla mancina, or from right to left, has never been printed.

On the reverse of the modern Title may be found an extract from the Life of Leon. da Vinci, by du Fresne, in which this volume is particularly mentioned.

It appears from the Title page (altho the name of the possessor has been obliterated) that it has belonged to Giuseppe Ghezzi, an eminent Painter at Rome; who, according to the inscrip "had obtained it, by a vast sum of money, to ~~perfect~~ complete the laborious collection of his library."

a very legible transcript of this original MS. is in the Collection at Holkham

W. Roscoe.

Flyleaf with nineteenth-century notes by William Roscoe and the signature of Thomas William Coke (see Introduction)

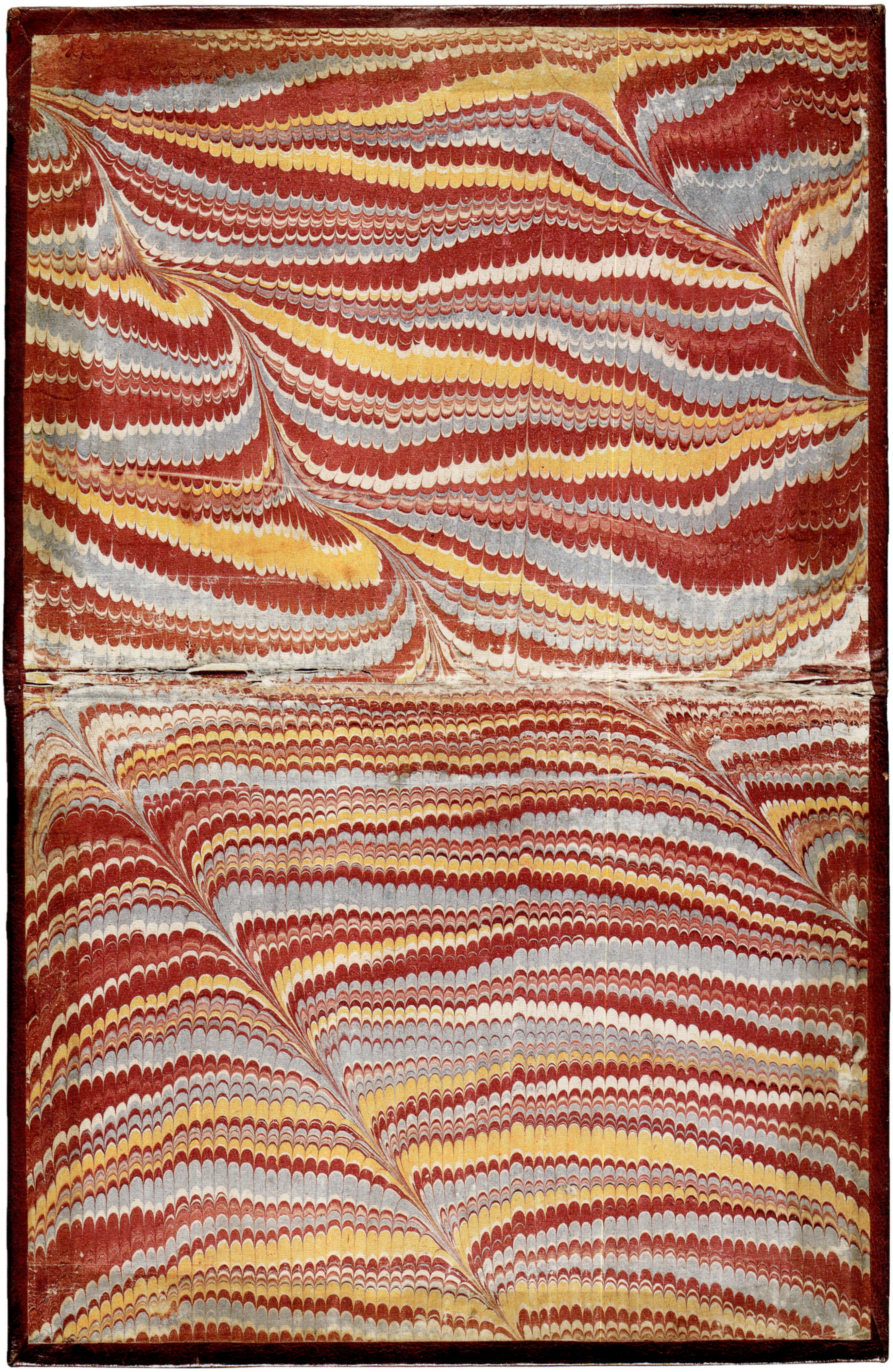

Inner side of eighteenth century cover

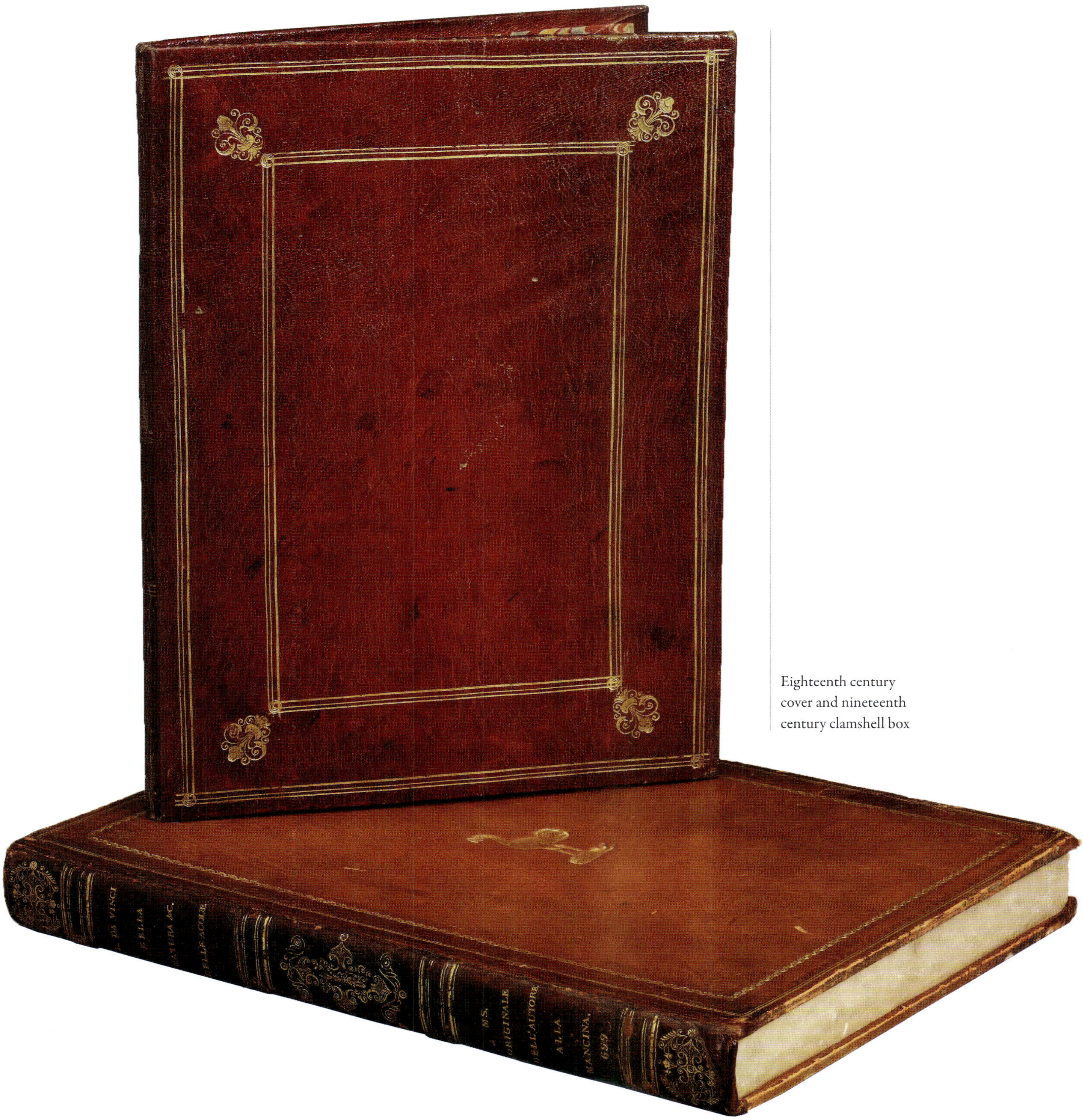

Eighteenth century
cover and nineteenth
century clamshell box

OXFORD
UNIVERSITY PRESS

Great Clarendon Street, Oxford, OX2 6DP,
United Kingdom

Oxford University Press is a department of the University of Oxford.
It furthers the University's objective of excellence in research, scholarship,
and education by publishing worldwide. Oxford is a registered trade mark of
Oxford University Press in the UK and in certain other countries

First Edition published in 2019

Published in the United States of America by Oxford University Press
198 Madison Avenue, New York, NY 10016, United States of America

British Library Cataloguing in Publication Data
Data available

Library of Congress Control Number: 2019935955

ISBN 978–0–19–883287–4

Printed in Great Britain by
Bell & Bain Ltd., Glasgow

The manufacturer's authorised representative in the EU for product safety is Oxford University Press España S.A. of El Parque Empresarial San Fernando de Henares, Avenida de Castilla, 2 – 28830 Madrid (www.oup.es/en or product.safety@oup.com). OUP España S.A. also acts as importer into Spain of products made by the manufacturer.